Sybil Lockhart

Mother
in the
Middle

⌇

*A Biologist's Story
of Caring for
Parent and Child*

A TOUCHSTONE BOOK
Published by Simon & Schuster
New York   London   Toronto   Sydney

Touchstone
A Division of Simon & Schuster, Inc.
1230 Avenue of the Americas
New York, NY 10020

First Touchstone hardcover edition February 2009

TOUCHSTONE and colophon are registered trademarks of Simon & Schuster, Inc.

For information about special discounts for bulk purchases,
please contact Simon & Schuster Special Sales at 1-800-456-6798
or business@simonandschuster.com.

Designed by Jan Pisciotta

Manufactured in the United States of America

1   3   5   7   9   10   8   6   4   2

Library of Congress Cataloging-in-Publication Data
Lockhart, Sybil.
Mother in the middle : a biologist's story of caring
for parent and child / by Sybil Lockhart.
p.   cm.
1. Lockhart, Sybil. 2. Women biologists—United States—Biography.
3. Pregnant women—United States—Biography. 4. Caregivers—United
States—Biography. 5. Lockhart, Ruth. 6. Alzheimer's disease—Patients—United
States—Biography. 7. Fetal brain—Growth. 8. Brain—Degeneration. I. Title.
QH31.L692 A3 2009
570.92—dc22
2008013490

ISBN-13: 978-1-4165-4155-4
ISBN-10:   1-4165-4155-1

*This book is dedicated to the memory of
Ruth and Robert Lockhart.*

THANKS

For sincere and thorough support of my writing, I thank: the Motherlode Writers: Marian Berges, Caroline Grant, Ursula Goulet, Rebecca Kaminksy, Sarah Kilts, and Sophia Raday, whose sharp minds and big hearts have carried me through every stage of this book; Amy Hudock and the old San Francisco Writers crowd who got me started; my Monday Gourdheads: Kathy Briccetti, Veronica Chater, Lynn Goodwin, Annie Kaooof, Suzanne La Fetra, and Rachel Sarah, whose raw honesty, quiet sighs, and tasty Pyramidal meals have sustained me; my sweet, patient editor, Trish Todd, and her cohorts at Touchstone; and my talented, kind, and relentlessly thorough agent, Linda Loewenthal. Suzanne La Fetra, thank you for a retreat like no other—a part of my soul lives on that hill, and no one understood me like you did this year. For moral support, loyalty, honesty, and friendship, I thank my sister, Alice; I love you. Thank you, Leah Korican, for thorough feedback, silly humor, artistic inspiration, and such undying confidence in me—and for being my best friend. For intellectual stimulation, friendship, and support of our love story, I thank Muffie and Scott Waterman and their young associates, Eva and Oscar. For the ongoing exchange of words, songs, comfort, and family, thank you, Polly Pagenhart, Jennifer Boesing, Amazing Maclain, and Sweet Jonah. Thank you all the mamas at literarymama.com for collaboration, support, and a super Web site; Ericka Lutz, Ted Weinstein, Jill Rothenberg, Andi Buchanan, and Heidi Raykeil for early advice; Michelle Withers, Bob De-fandorf, Leah, and Polly for help with the seed essay; Susie and Hugh for grassroots promotion and encouragement; Allyson and Leslie for taking the kids while I worked; Joanne Catz-Hartman for pep talks; Ken Frankel for *Science* and *Nature* delivered to my porch; Jessica Pisano for reading my mind in the lab and ever since; the Dana Foundation for accessible brain news; the UC Davis Alzheimer's Disease Center for evaluation, advice, and support; Billy Kenney for Brainworks and other tidbits; William Bridges for his book *Transitions;* Ira Black for *The Changing Brain;* Natalie Angier for *Woman: An Intimate Geography;* Jonathan Franzen for his thoughtful essay about his father. Thank you, Tim H., for Pema. To my scientific mentors, Susan Birren, Irwin Levitan, Eve Marder, Mikey Nusbaum, Claudio Pikielny, Barry Rothman, and Gina Turrigiano, thank you for supporting and teaching me; I admire you so much. All my Noonies past and present and your associates worldwide: nothing happens for me without you. And most of all, I thank you, Pat, Zoë, and Cleo, for seeing me through this; I love you all the way to bursting.

CONTENTS

Thanks   vii

1. Friday Night   1

2. Blue Moon   9

3. Telephone   21

4. In Utero in the Lab   29

5. Hatchling   43

6. Inklings   57

7. Leaving   77

8. Fall Back   89

9. Vertigo   109

10. Pivot   131

11. Diagnosis   143

12. Impulse Control   161

13. Baggage   179

14. Number Two   189

15. Regrouping   199

16. Fourth of July   211

17. Take Care   227

18. Mama in the Middle   241

19. Lost and Found   261

20. Days of the Dead   277

21. Writing Home   293

CHAPTER 1

Friday Night

Dinner is over. This is the windup period before bedtime. My husband, Patrick, is working late. I'm trying to load the dishwasher while five-year-old Zoë spins in circles on the black-and-white tile floor holding her fairy wand, and one-year-old Cleo shuffles back and forth in Patrick's black loafers, giggling.

"Five minutes, girls," I warn, glancing up at the wall clock over the sink. It's 8:35.

Ma, still seated at the kitchen table, looks down at her lap, bending forward so I can see the hump of her back under her favorite red plaid flannel shirt. Her gray hair is greasy and thin and falls only to her shoulders, but looking at her, I remember her hair of many years before, long and dark. This lesser hair seems gentler somehow—a quiet, softened version. And Ma, once the head of my childhood household, the breadwinner of the family, and the coordinator of all our activities, has softened, too, softened and receded. She has difficulty initiating conversation; she can't find the courage, or she can't find the words; I'm not sure which. She frets and hovers, unsure of herself and afraid to ask for help.

Clasping her hands together now, Ma looks up with a plaintive, pained tightening of the skin around her eyes and opens her mouth to speak.

"No. Never mind." She shakes her head and looks back down.

"What, Ma?"

"No, nothing. I'm just being stupid. Never mind me."

I step forward, laying my hand on Ma's shoulder. "Ma. We try not to use that word around the kids, okay?" I feel her wince, and I quickly add, "Plus, it's not true."

"Oh, I knew that. That was stupid. No, I mean—not stupid!" She glances quickly in Zoë's direction. "Never mind. I'll just shut up."

I can't decide whether to smile or scowl. "Ma," I ask, with as much patience as I can muster, "what is it?"

But I know; I know it by heart, because she stays over most Friday nights now, and it's the same every week. She wants the TV on, but the remote control confuses her. She needs the bed made up, but she doesn't understand how the futon couch unfolds. And she expects to be in bed by nine. If I go up to bathe the kids and don't get back down in time, she'll stand at the foot of the stairs, sighing dramatically and wondering out loud what's keeping me. She'll fretfully pull at the futon but stop herself and again glance up the stairs. If I tuck in the kids and walk back downstairs at nine-fifteen instead of nine, it may as well be two a.m.; the waiting and the uncertainty of those fifteen minutes weigh on her that heavily.

The problem for Ma is that if she doesn't do everything just so, something might slip by her. It could be something small, like forgetting to brush her teeth. But it could also be a stove burner left on, tall hot flames licking out into the dark kitchen for hours as she sleeps. It could be the door left standing wide-open with the black night yawning in. Or maybe her whole world will dissolve. The truth: it is dissolving, right there inside her head. Sticky mats of neurons slowly clot there, gradually obscuring patches of her own life from view. Some of her perceptions remain clear; others are very slowly melting into a murky blur. And she doesn't know what will go next.

The schedule saves her. Clean and sure, it keeps her from wondering. Her list is intact, and she checks it off: *6 p.m. dinner. 8 p.m. television. 9 p.m. bathroom & pills. Then read in bed.* If the bed is not made up well before nine, something feels askew. She doesn't quite understand, but she feels out of control. Maybe she has missed something important. Maybe she's done something wrong. So she lives in a constant state of low-level worry that swells into a panic as the clock ticks past the hour.

Zoe zaps Cleo with her wand, eliciting a gleeful toddler shriek. She strikes again, but too hard, and Cleo bawls. Remorseful, Zoë tries to soothe her by picking her up, hauling her to the couch, and kissing her ferociously. "It's okay! It's okay!" she insists, as Cleo begins to cry in earnest. I look feebly from my two kids, now crying together in long, wavering wails, to Ma's anxious hands, wringing each other in her lap. The distress Ma projects when her schedule is disrupted infects me immediately. As she begins to fret, so do I. I have to soothe her. I have to make it stop. I must get these kids into bed, too, but if I don't take care of Ma this instant, my own world may dissolve. And in a way I hate her for it; I hate her for making this helpless, slow descent into the darkness.

Two years ago I thought my life was complex. I worked at UC Berkeley running an undergraduate neurobiology lab class, and at home I took care of my toddler. Between work, child, marriage, and friends, there never seemed to be a free moment. Now I see how easy I had it. My weekends were free, I had only one dependent, and the neurobiology stayed neatly confined in incubators and under microscopes. When I locked up the lab at five-thirty to go home, my life was essentially my own. Now at home I am never free, and I'm continually bombarded by real-life neurobiological phenomena. I spent five years in graduate school and another three in a postdoctoral fellowship, learning all about the brain, but in taking care of my family I find that I'm more preoccupied with the workings of the mind and nervous system than ever before.

As mother to a toddler and a kindergartner, I am so used to being able to fix things, to nourish. I kiss the boo-boos. I serve the food. I get up to comfort them when they cry in the night. And I see them heal, learn, and grow; they rise up into the light, gathering knowledge and developing insight. As they play and think and create, becoming clearer and brighter, I can almost see the myriad intricate synaptic connections being sculpted and refined within them. Meanwhile Ma seems to wilt and fade, as sticky protein oozes into her synapses and the fibers that structure her neurons begin to crimp and tangle. The girls' development puts Ma's dementia in dramatic relief. I think I am less tolerant of her mental deceleration because I'm simultaneously witness to the tornado of their intellectual development, and because I get to participate in their progress. No matter how much I give to Ma, she only becomes smaller. Ever so slowly she loses comprehension of the world, and as she does, she loses pieces of the person she once was to me. That is how it works, that is how she incites my empathy and my rage in equal parts. I should be able to fix it. I can't.

All though my childhood, Ma was the provider. Daddy was legally blind and received disability from the county. She made most of the money—and she did all of the cooking. She taught us to read, and she took us for walks in the hills, where she shared her love of the outdoors. On Friday nights my parents' friends gathered for guitar playing, puns, political discussion, and drinks.

While the kids ran wild, Ma served smoked oysters; she made the martinis. She was strong-bodied and had strong, simple principles. Care for your loved ones. Eat well. Enjoy sex. Work for peace.

Ma was warm and welcoming. She cared about animals, about her first-grade students, about the planet. She told me that the most important thing in life was love: finding someone to love, finding someone to love you. I know this to be true, and I repeat it to my own children. Ma also believed in plain words. When the county jail down the road was rebuilt, the new sign said "Detention Center." For three mornings after that sign went up, Ma walked the dogs down to the new building at six a.m. and tacked a banner over the new sign. "It's a JAIL," it read, and in smaller script at the bottom, "Detention Center indeed!" She lettered her banner in the same beautiful calligraphy she used to make our Christmas cards and the leaflets she handed out at peace rallies in San Francisco.

With every passing day, Ma remembers less of our history together, whereas the layers of meaning and memory seem to keep deepening for me. That's true tonight, as I think about bedtime in my childhood. It was complicated by the fact that my sister and I essentially occupied our own flat at night. The house we lived in was on a hill. It had an upstairs and a downstairs, but the two were not connected by an indoor stairway. When I was a baby, we all lived on the ground floor, which was the top floor, measuring in at about five hundred square feet, and this cozy arrangement suited me just fine. But before I had stopped wetting the bed, my parents decided to use our bedroom instead as a library and study. They moved Alice and me to the two small downstairs bedrooms. Every evening we made the journey down the side stairs, past a hedge of sweet-smelling oleanders, along a cement path that ran the length of the house, and then through the enclosed porch that served as my father's workshop, to our two little rooms at the bottom of the house. We turned on all the lights as we passed through the workshop, to chase away the fearsome shadows cast by Daddy's power tools, and then sprinted for our beds.

In winter and spring, when we rose before daylight and dashed barefoot through the rain to the warmth and safety of Upstairs, it was impossible not to step on the snails. I felt the wet crunch as I ran up the slippery steps, and then there I was in the dining room, with wet hair and freshly dead, slimy snail innards all over my bare feet. I would frantically wipe the slime off onto the burgundy shag rug, reaching for a tissue to get the last bits out from between my toes. There was always coffee brewing. The smell of coffee meant I had survived another night down there; it meant I was back with my parents. I was safe. To this day that smell still signifies the comfort of Mommy, of Home.

I love passing on the snail-crunching lore of my childhood to five-year-old Zoë. She squirms, delighted and horrified to imagine it. But she asks me, "Why didn't you have slippers on? Why didn't Aunt Alice and you get to sleep upstairs anymore? What if you needed your parents in the night?"

These questions remind me of how I raced up and down those stairs, sure that the hands of child-grabbers were reaching out to catch at my ankles in the dark; how I ran past those looming shadows cast by Daddy's table saw and workbench on the porch. My parents didn't seem to understand that the locks on the doors downstairs couldn't possibly keep the scary things out; that there was no way for my sister to save me, since she was just a kid, too, and I knew Daddy and Ma wouldn't hear me screaming down there. For over a year, back when Ma still came down to tuck us in, I cried for her after she had gone back upstairs, and I hiccupped myself to sleep. It just didn't feel right to be so far away from her—from Daddy, too, but especially from her.

The desperate longing I felt those dark nights downstairs informs me now as a parent and as a caregiver: this is what you don't do. You don't leave them not knowing where you are. You don't leave them scared and lonely on another floor of the house with the darkness and the rain and the slimy snails between you. Maybe this is why, when my turn arrives, as forty-year-old daughter-turned-mother, I can't just leave Ma downstairs, worried and waiting. Instead I ask Zoë for help.

"Sweetie, can you please read Cleo a little story while I help Gram get ready for bed?"

Zoë's face lights up with big-sisterly responsibility, and Cleo swipes at a tear, looking up expectantly. They make it all so much better sometimes, for me and for each other. I want to stop and scoop them up in my arms to thank them; I want to nuzzle them and lose myself in their elastic, happy warmth—but I know I've just bought myself a maximum of one and a half minutes. I lead Ma to the futon in the living room, which doubles now as her room most weekends, and together we perform the ritual of arranging the five blankets and two pillows in exactly the same order as we do every Friday night. This is an odd feature of Ma's condition; as she forgets events of increasing importance, the simple, quotidian routines like laying blankets on a bed have taken on a disproportionate importance for her. In her fear of losing track of things, she has begun to guard the repeating patterns in life with a fierce, almost superstitious insistence: the blue blanket must go on top; the little red woolen throw must be at her feet. I wonder at her ability to recall this arbitrary sequence of colors and textures, given the way she recently got lost returning from the restroom in our favorite café.

I glance slyly at her.

"So, you want the blue one on the bottom, right?"

I'm teasing, partly to test if she still has her sense of humor, and partly because I've gone too long today without adult company and need a laugh. She looks up sharply, alarmed, but when she sees my smile, she returns it.

"Ah, you had me there for a second!" She laughs.

I'm gratified and relieved to have conjured this flicker of what I think of as the "real" Ma, and I grin as I tune the TV to the channel for her favorite program, *Providence*. I bring her the remote control and carefully point out the mute and off buttons to her. Finally, setting down her cup of Sleepytime tea, I kiss her soft cheek. I'm trying not to breathe in too much of her odd scent. My father died almost twenty years ago, and still it seems to me that her clothes smell faintly of his cigarettes, mingled with the odors of cooking oil and mildew. I tell her, "Tomorrow we'll go to Saul's for breakfast," and impulsively, I hug her tightly.

"Thank you," she says. She has tears in her eyes. My spontaneous warmth mixes with something like embarrassment, but much sadder. I don't *like* having to take care of her. When she's grateful like that, I feel like an impostor. "Don't you thank me, Ma," I whisper. Then I turn and herd the kids upstairs.

There the air smells sweeter and the lights seem warmer. After I change Cleo's diaper, I pause to bury my nose in her soft belly, and she giggles with pleasure. I snuffle, breathing in the warm, sweet goodness of her skin. Patrick, finally home from work, comes upstairs and leans in the doorway, telling me about his day, while I help Cleo into her pajamas. Wild and naked, Zoë bounces on the bed, and I turn to admire her lean stomach and legs. I heave a sigh of relief. We are a safe distance from Old; we are upstairs, together, healthy, and sane.

CHAPTER 2

Blue Moon

I met Patrick in 1993 on a volleyball court in Waltham, Massachusetts. We were both in graduate school at Brandeis University, he in computer science and I in cell and molecular biology. The night we met, he was wearing army surplus shorts and a Frank Zappa T-shirt; his long brown hair was pulled back into a neat ponytail, and his skin was utterly un-tan. I noticed his lean body and admired the way he passed the ball to every team-mate, even the weakest player. I thought, Ah, he's a cooperative, sociable computer guy. His eyes twinkled as he firmly shook my hand after the game. "I'll bet he wears Birkenstocks and likes children," I told my friend Marco as we left the courts. This proved to be an accurate assessment.

Pat and I both led simple, circumscribed lives: a few friends, a lot of work, and Friday night games at the gym. He made the first move, in an e-mail that arrived the day after we'd met.

"Sybil, is that you? It's Pat, from volleyball. Nerdy guy, long hair." Hah. A tingle of pleasure ran through me as I realized a part of me had been thinking about him and that long hair as I worked in the lab. We carried on a pleasant e-mail banter all that week. He sent me jokes, anecdotes, tidbits he'd found on Usenet groups.

"Is your name Pat or Patrick?"

"Yes."

"Okay, then, Patrick. I like Patrick best."

"Fine, but don't tell Pat; he'll be crushed."

We courted via e-mail initially because I spent far too much time in the laboratory and he was at his computer about ten hours a day. I was studying a novel gene I'd helped to clone from the nervous system of a sea slug. Pat was working on a way for computers to write their own programs using "genetic programming," a technique loosely based on evolution. For the first two weeks we wrote e-mails every day and met in person only on the volleyball court. Over the next couple of months we branched out, for a fall hike in the New Hampshire woods, a tour of MIT's steam tunnels, a blues show at the Sittin' Bull pub. Eventually, one night during a commercial break in *Saturday Night Live,* he told me he'd like to kiss me. "That's good," I said, "I want you to," and we went from there.

I had moved to Massachusetts from California four years earlier, but my heart was still back in the Bay Area. My best friends were still there, and my mom still lived in the house I grew up in. I called her every weekend, we sent birthday cards and Easter greetings, and I visited her and her boyfriend, Eddie, every year for Christmas or Thanksgiving. Often my sister, Alice, and her husband, Josh, also showed up for holidays, and most years Ma rented a shabby little cabin in Stinson Beach for Christmas. So it was that a few months after that first kiss, I took Patrick to the beach to meet my family.

We had a simple, relaxed routine at the cabin: we took long walks on the beach with Ma's and Alice's dogs, ate and drank rather decadently, and read as much as possible. I usually invited a friend or two out to join us, and Ma had always made my friends welcome, but we were all gun-shy after my last serious relationship, which had lasted five years and had begun and ended in betrayal. Since then Ma still greeted my dates warmly, but she held something in reserve, almost as though she had been the heartbroken one. "You be careful," she told me every time I mentioned a new love interest. "You do what you want to do, of course, but I'd hate to see you hurt again." I was trying not to worry about what she would think of Pat, precisely because I already felt strongly about him myself, but I needn't have worried. Somehow he managed to make it onto Ma's good list on our very first morning in the cabin.

That morning, before any of the others were up, Ma and I sat at the kitchen table drinking coffee. Alice and Josh, Ma and Eddie had the back rooms, and Pat and I were sleeping on the living room floor, where Pat still lay sprawled and softly snoring. Ma read and I listened to the waves massage the shore in the distance. Ma and I were both morning people, and although we didn't talk much, we both understood that this quiet time in the early hours was our special alone time. I was just about ready to get up for a walk along the empty beach when Patrick rolled over with a sigh in our mess of sleeping bags and pillows at the other end of the room. Ma's golden retriever, Nita, who'd been sleeping at her feet, lifted her head, and her tail began to thump the floor.

"Uh-oh," I said, looking up with a smile, and I nudged Ma with my elbow.

Doing the happy dance of an eager puppy, Nita clicked across the parquet floor, her tail low but softly wagging, and gave Patrick's nose a tentative lick. Ma and I stifled a giggle, waiting to see what would happen next. Pat just lay there until Nita offered another, more vigorous kiss, at which point he groaned. But then, instead of rolling over and burying his head under his pillow as I would have done, from under a swath of long hair Pat emitted a gentle chuckle:

"Well, good morning to you, too!" He rolled onto his back. "Okay, okay, come on over." And then he just lay there, with a big grin on his face, while Nita licked his cheeks, neck, and forehead, wagging all the while.

Ma watched all this with a hand pressed to her mouth, and then she leaned over to me. Her voice was full of tenderness and wonder.

"That Patrick is a good man."

Within a year Patrick and I were engaged and living together in a small flat across town from Brandeis. Just after I defended my Ph.D. thesis, we sat down to puzzle out the logistics of our new life together. This was somewhat tricky. Although we had been in the same year of graduate school when we'd met, I had been out in the workplace for a few years before coming to Brandeis, whereas Pat had gone straight through; he was a full six years younger, and our needs were different: I was almost ready to start a family, but he was still focused on finishing a dissertation. We were no dummies, though. Aware of the potential complications, we responsibly locked ourselves into a university conference room with a wall of whiteboards and a box of dry-erase markers to brainstorm it out. An hour and a half later we emerged with a lovely plan, a carefully ordered calendar of events. He would take six more months to finish his degree while I did a short post-doctoral fellowship at Brandeis to wait him out. Then we would marry, move to California to find jobs in academia and technology, and start a family.

It was supposed to go in that order, but then real life took over. Patrick's thesis adviser left the university to manage a start-up company in New Jersey. Pat's work stalled out, his six months growing to nine and then twelve. Our time line constricted month by month until his thesis deadline, the date we'd decided to start trying to conceive a child, and our wedding had all converged. Nineteen ninety-six was going to be a big year for us. And we didn't know it at the time, but the changes taking place inside Ma were about to add one more factor to the mix.

During our week at the beach, Ma was still Ma. She may have been quiet Ma, less energetic than in years past, but she was still fully herself. My earliest conscious awareness that she was subtly changing would come two years later, the week of my June wedding to Patrick, by which point it seemed everything was happening at once.

We had intended to exchange vows on a grassy Vermont hillside, but that morning we awoke to a light summer drizzle, so we relocated the ceremony to a small white country church down the road. Snapshots from that morning show my bridesmaids in a dim bedroom, brushing and braiding my long hair and then Patrick's; Patrick in a white dress shirt with a mandarin collar, bending to kiss my shoulder; the belly of a cloud lowering itself over the lush green hilltop. In later photos, the two of us stroll through the mist to our reception, hand in hand under a dark green umbrella, and still later, with the sun finally out, children trail balloons across the outdoor dance floor and women in long dresses toss a NERF football in a huge green field. In one picture, our mothers sit together on a couch at the local inn. My new mother-in-law, Susie, is hearty and authoritative in low heels and lipstick, while Ma, tastefully dressed but gray-haired and small, turns an overwhelmed, teary-eyed smile to the camera.

I felt protective of Ma that day. She seemed awkward, vulnerable. Eddie hadn't come with her, for reasons I had paid little attention to. To me, Ma's boyfriend seemed mere dressing, whereas my mother was an essential ingredient for the wedding, but when I saw her, I was struck by how small she seemed, there alone. She had flown out from California and driven in from the airport late the night before, looking wide-eyed and ragged. I hadn't seen her in several months, and I noticed right away that she'd lost a lot of weight. There was a frailty about her. *She's practically a senior citizen now,* I thought. Her sixty-fifth birthday was a month away. *I wonder if she's depressed.* My dad was now ten years dead of lung cancer. My godmother, Eleanor, had just been diagnosed with the same disease. Eleanor and her husband, Daniel, were Ma's closest living friends. Maybe Ma was suffering from cumulative sadness.

"It was beautiful, darling—it was just perfect," Ma murmured in my ear as I hugged her after the ceremony. She kissed my cheek, and then turned to Patrick. "Patrick, I know you two will be good and love each other, and that is the most important thing of all."

I felt Pat's arm tighten around me. "We will take good care of each other, Ruthie," he said. I bit my lip, trying not to cry again.

"Ahhh, here they are," a voice said from behind me. "And this must be Mrs. Lockhart." Patrick's aunt reached a hand out to Ma, who suddenly looked bewildered. "So glad to meet you. I'm Patrick's aunt Diane—and this is his uncle Richard." Pat's uncle had stepped up to shake his hand and now turned to Ma. "Dick Lynch. Very good to meet you."

"Oh!" Ma said, looking uncertain of herself. "Hello. I'm Ruthie. Sybil is my daughter." Ma pronounced the words in a loud voice, as though speaking to a roomful of first-graders. I chuckled a little uneasily.

"Well, you must be so proud today," said Diane.

"Yes, I suppose I am quite proud," Ma said, and she turned

to me. I waited for her to say more, but she just stood there and continued to smile intently. Her face reddened. There was a momentary uneasy pause, but then more guests came to greet us, and as the conversation picked up again, Ma moved to sit down with my sister, Alice.

Had she always been so awkward, so shy in a crowd? I wondered, but soon our band, the Blues Doctors (made up entirely of M.D.'s and Ph.D.'s), began to play loud, fast dance tunes, and Patrick and I and Alice and Josh moved out onto the floor, leaving Ma rooted to her folding chair. I can still see her there, with her wide, watery eyes that alternately glanced shyly around the big tent and stared down at her lap. Watching her, I felt a twinge of bittersweet longing. This day formalized a change in our relationship that had begun decades ago, as I had shifted into adulthood; I was now officially grown, no longer her dependent. I wanted to soothe her, to thank her, to start a round of applause for this woman who had taught me to love—but she shied away from the attention, embarrassed. As others got up to dance, I watched Ma withdraw to a seat away from the crowd. Although she did eventually warm and join us for one or two dances, as I watched her in that moment of retreat, I felt regret trickle in to dilute the excitement of the day. It was only a vague, nagging feeling, but somehow, as I gained a husband and a new life, Ma seemed to be losing something—losing not only me but also maybe a part of herself.

It was still too early to know that Ma was experiencing the first signs of dementia. Even before this day, before a hint of a symptom had emerged, some of the cells of her brain had begun to disintegrate and die. Alzheimer's is a sneaky disease. It starts out subtly and can grow into something monstrous, but its progress is maddeningly slow, a gradual reduction of a whole person to increasingly fragmented versions of herself.

It begins in a few nerve cells near the center of the brain, where memories are stored. Early on it is clinically undetectable, but inside, cells are already changing. By the time I married Patrick, if we could have spied into certain regions of Ma's brain— her hippocampus and cerebral cortex—we would almost certainly have found that they already had the hallmarks of the disease: twisted filamentous tangles of one protein inside the nerve cells, and clumps of another, fragmented protein accumulating in the spaces between them. The tiny cargo sacs of enzyme and other cellular components in transit from one point to another within her nerve cells may already have been stalled out and piling up, unable to move along their normal paths. Communication between neurons was also probably already beginning to wane, and we might have observed the seemingly paradoxical birth of new nerve cells amid this inchoate chaos, as her brain delivered new recruits in an apparent effort to stave off the impending crisis.

The affected cells were used to process, store, and retrieve information; Ma's memory and her ability to think depended on them. But at my wedding, even though some may have been beginning to fail, she had such an abundance of them still, and so few were damaged, that her behavior did not deviate radically, and no one, not even her daughter the neurobiologist, saw her shyness and fatigue as indications of trouble in her brain.

In truth I may have been a bit distracted that day, not only because I was getting married but because I had my own biological process to consider. Just as, one at a time, cells in Ma's brain had begun to founder, certain healthy cells deep within my own body had just now begun to bud and multiply. I'd been told that rain on your wedding day was a sign of fertility, but that gray morning I had a strong hunch that we were already fine on that count.

I believe in free will; I do not subscribe to the ticking-clock notion that *every* sexual move we make is dictated by hormones.

Yet at thirty-four, I was well into my childbearing years, and I was an unabashedly biologically driven soul. Every month without fail I entered a period of midcycle frenzy, when I found Patrick insanely attractive. So it had been one night just days earlier, when he had returned home late and quietly crawled into bed with me. We had intended to start trying *after* we were married, but there had been a change of plans.

I had arrived home from the lab not long before him, but I had already entered semiconsciousness, half-dreaming of the neuronal growth hormones I'd been reading about that afternoon. It was late June, and even downtrodden little Waltham felt luxurious and silky with the warmth of early summer. The air had softened, leaves rustled gently in the trees, and I had kept the windows open that night so the perfect air could move slowly though the room as I slept. My muscles felt energized, my skin smooth and elastic; I craved movement and contact. In less than a week we would be surrounded by our friends and family, exchanging vows. I wanted to savor these last few nights of being just lovers. I also thrilled at the prospect of unprotected sex and the vast unknown territory of conception, children, family. I stood on a high, airy cliff, my arms spread wide, ready to leap, and as Patrick sneaked under the covers and softly kissed me, the monthly tide of hormones swept in.

I surrendered immediately, absolutely. Clothes became intolerable and touch imperative. As I closed my eyes and kissed him, multiple chemical messages urged me along toward one goal: *make a baby*. My ovaries were already dumping far higher than the normal doses of testosterone and estrogen into my blood, and the touch of his lips on mine, his long hair sweeping across my face, my chest, my belly, caused another hormone, oxytocin, to rush in as well. Oxytocin, molecule of touch, fosters affection in its subject for the person who causes it to be released. It wells up in the mama as she nurses her baby, and it causes uterine contractions not only during childbirth but also during orgasm. Oxytocin flowed through my veins and also through Pat's, immediately working to heighten both our sensitivity to touch and our desire for touch, in a delicious expansive spiral of positive feedback.

Just for a moment before I tore off my pajamas, I was conscious of this hormonal cocktail swirling in my veins, of the way each influenced the other, with our caresses releasing the oxytocin, and estrogen in turn enhancing its effects. It may have been the estrogen itself that infused me with clarity, conferring the crystalline vision and perfect energy required to actually reason amid such heated confusion. Estrogen would have kept my senses of smell and sight sharp as I paused just long enough to think very clearly, *We could really make a baby,* just long enough to say "Condom?" to Patrick, and hear his response, "Why bother," before the spike of oxytocin pushed me into full delirium. Then, as my nipples and clitoris engorged with blood, becoming taut and sensitive, testosterone like a whip at my back drove me toward the goal, that pinnacle moment of deep connection, every muscle of my body driven to writhe and thrust. I was an unstoppable, intelligent force, and together with Patrick, I entered a furious crescendo of motion, heat, elation, and gratification.

Afterward, as I lay smiling in the dark, my uterine muscles continued to pulse reflexively, and I knew that with each delicious involuntary contraction my cervix was dipping down into the sperm and sipping, actively ushering many millions of little swimmers up and in. They would make the dark journey through my womb and into the mouths of my fallopian tubes, just as a mature egg was swept softly down by gently waving cilia to greet them. I could almost feel it.

Sometime after our honeymoon, a midwife charting the time course of my pregnancy would note the approximate date of conception as our wedding day, but I knew better; I was certain that a minuscule dot of a new life had attended our ceremony along with the other guests in that cozy church, had been there with us on our stroll through the summer mist and under the white tent where we held our reception: a tiny ball of cells, dancing in my belly as I danced with Patrick, whirring with the momentum of new beginnings.

We had placed disposable cameras on all the tables at our reception, and someone snapped one picture of Ma, my sister, Alice, and me, alone in a sea of chairs. We are all laughing. Ma's head tilts back, her blue eyes bright with mirth, and Alice and I sit close by on either side looking contentedly jolly. I gave that photo a prominent spot in our wedding album. Maybe I intuited that this was a moment to savor and cherish. I didn't realize it then, but the timing was uncanny: two dramatic biological forces seemed to be set in motion almost at the same moment, one inside of me and one inside of Ma, each almost the inverse of the other, and together they were about to propel us into a future that I had not imagined for myself.

That night in our room at the inn, Patrick slept first, and I lay contentedly in bed among the rose petals scattered earlier by our friends, my head filled with happy scenes from the day. I stared up through the open window at a mostly full moon partly masked by slowly passing clouds. I didn't know it at the time, but that was a blue moon, the second full moon in one month. Waxing or waning, I wondered groggily, and then I was asleep.

CHAPTER 3

Telephone

Eight months after our wedding on a Saturday morning, I stood at the window of our little apartment and stared resentfully out at snow-covered cars under a bleak gray sky. I needed to go in to the lab and check on the cell cultures I had put in the incubator Friday afternoon, but I was overwhelmed by my own convexity: in my dad's old red-and-white-striped pajama bottoms, with my innie belly button now pushed into an outie, I was a hot-air balloon; I was a great striped melon with one giant, complex seed thumping around inside me. Prenatal aerobics class had kept me fit, but I could not bring myself to waddle out there and shovel the steps.

Instead, I lowered myself onto the couch and dialed Ma's number. I had called her just the night before, but suddenly I wanted to hear her voice again; I needed someone who would empathize. I could picture her standing in the sunny kitchen of the California home where I grew up. She would be pouring herself a bowl of Shredded Wheat and a glass of grapefruit juice. Eddie, if he'd spent the night, would be at the dining room table with a cup of coffee. It was already spring there, and the hummingbirds would be hovering at the feeder in the front yard. I realized that Ma hadn't sent her usual late-winter box of Meyer lemons from her garden. When they came, I could sit at my desk, with the clack of the neighbor kids' hockey puck sounding on the icy river out back, and sip fresh lemonade, just to remind myself that spring really would eventually come.

"Hello?" She sounded breathless.

"Hey, Ma. Just thought I'd call and say hi. Did I catch you at a bad time?"

"Oh, Sybil, thank goodness. I thought it was going to be one of those thingamajigs—people wanting me to buy things. I wish they'd leave me alone."

"Telemarketers?" I laughed. "They think you're old and gullible; let 'em know you're on to them and you're not buying it, and they'll leave you alone."

"Oh well, I suppose so."

"Well, you probably didn't expect another call so soon, but I just missed you all of a sudden."

"What do you mean? We haven't talked recently, have we?"

I inhaled sharply, realizing she'd forgotten our entire conversation of the evening before. This wasn't the first time, and if I called after she and Eddie had already had cocktails—which they did most evenings—it was particularly likely to happen. I slowly let out my breath. My mom had what I called a "dedicated drinking habit." Which was my nonconfrontational way of framing my suspicion that she was an alcoholic. I had never felt qualified to judge. She loved her drinks, and she had for years. And we all drank—alcohol was just an integral part of our family culture. Now, though, I was feeling more qualified. But we could barely talk anymore! The forgotten conversations were beginning to stack up; this would take its toll on her health if she continued.

Dementia was not even on my radar; I knew she was far too young for that. I was just thinking maybe it was time to gently suggest she cut back on the booze.

I had been worried about Eddie and his drinking for some time as well. Several years earlier Eddie had been in a severe car crash in which his then-wife was killed. Since then he had suffered a series of strokes that had left him with a frustrating language deficit—not to mention his broken heart—and he drank . . . like a drunk. On one weeklong trip to Stinson Beach, Patrick and I had stumbled upon two *gallons* of bourbon stashed in Ma and Eddie's closet. Morning or night, there was always something lacing Eddie's drink. Ma loved him, and she also loved her drinks, and that was her right. But if Ma and Eddie were drinking so much that she couldn't even remember a conversation with her daughter the day after, that was a problem.

This particular conversation had been important to me. Susie, my widowed mother-in-law, was planning to remarry in Connecticut just after the baby came. I was hoping Ma would come for a visit and help out while Pat was at the wedding.

"Ma, I called you last night. We talked about Susie's wedding?"

"Huh. No, I must have been tired."

Tired. I paused as that sank in. I wondered if Alice was having the same trouble talking with Ma in the evenings.

"Hey, has Alice called you lately?"

"I called her this morning, but Josh said she went out to buy bigger maternity clothes," said Ma. "She's going to call later."

As it happened, Alice was also pregnant. In fact, she had probably conceived the same week I did, possibly even the same day; our pregnancies had overlapped perfectly, down to our identical March due date. We'd both had amniocentesis; I was having a girl, she a boy. I stared out the window and pictured Alice in her parka, walking big-bellied down a rainy Seattle street with her dog, Pato. In December, instead of going to our usual beach cabin, my family had gathered in Seattle for Christmas. When we had arrived, Alice and I had stood belly to belly for a Kodak moment, and I had glanced over at Ma, wondering how it must feel to see both her girls plump with babies of their own. I guess I expected her to cry and hug us a lot. But instead she hesitated on the sidelines, a passive observer, until I pulled her hand over and placed it on my belly, commanding, "Ma, *feel* this thing already!"

Now I missed both Ma and Alice. Being pregnant was making me more homesick than ever.

"I wish I could see you guys more. I wish I could just teleport over there right now. What's the weather like? We had six inches of snow last night, and it's still about twenty degrees out."

"Snow! Yesterday it was seventy-five degrees here."

"God, now I *really* wish I could be there."

I longed for that sweet, crisp California warmth. I wanted the ease of just being there. I wanted to sit at Ma's table and talk about babies. I wanted her to ask me how I was feeling, to empathize, woman to woman. I thought I'd go insane if my only resource continued to be those infernal *What to Expect* books. I'd had no practical baby experience, and I longed for guidance. I didn't know what to *do* with a baby. All I knew was how desperately I already cared about the one inside me, and in one short month I'd have to release her into the big, scary world.

"Well, honey, at least you can call. How are you feeling? Everything's okay with the baby?"

"Yeah, it's fine. Hey, Ma, did you nurse Alice and me?"

"Well, I think so."

"Well, what was it like? Did it hurt?"

"Oh, I don't know—I suppose it was fine. I don't recall it being a problem."

She didn't recall? Was that possible? Ma, who had been so determined to have babies that she had refused to marry Daddy until he agreed to children, couldn't remember a detail like that?

"Well, did you also feed us formula? Wasn't that real popular back then?"

"I must have, I guess, if everyone else was doing it. I don't really know. I wish I did, honey. I know you will do fine, though. I'm sure it will all come naturally."

I didn't know what to say. I wished she'd be a little more enthusiastic, and also offer more concrete advice. I recalled several years earlier the two of us going to a baby shower for my childhood friend Regina. Ma had surprised me by buying several beautiful, expensive-looking outfits in infant, baby, and toddler sizes, little booties, baby toys. I'd watched Ma more carefully than I had watched Regina that afternoon, struck by how much care she had put into the gifts, how genuinely excited she seemed at the prospect of this little creature. At the time I'd had no particular interest in babies; I hadn't understood all the fuss. But seeing Ma hold up a tiny purple union suit as it was passed around the circle of chattering women in Regina's living room, I'd caught a glimpse of something, her nostalgia for a time in her life, maybe—or, I thought, perhaps her desire to be a grandma. What had happened to all that enthusiasm?

I wondered about Ma's new apathy, but I quickly told myself that maybe with the passing years, it was natural for the once-so-critical facts of early parenting to fall away. Later I would understand how common it is for family members to perform this kind of mental contortion to validate the behavior of the demented. Others have observed that this occurs in much the same way that we compensate for missing sensory data—the partially heard word, for example, or the incomplete pattern of a face. Our brains are expert at filling in the gaps, so the picture appears complete in spite of the missing pieces. Likewise, we fill in missing data in our relationships with ailing loved ones, inventing excuses, explanations for their mental deficits. Maybe she hadn't lost the enthusiasm, I told myself; maybe she had merely discarded the trivial details. That seemed practical enough, I thought; I'd probably do the same when I was sixty-five. Only my body knew better; somewhere up in my sternum, there was a familiar tightening, a rising alarm.

"So how is Eddie doing?"

"Oh, he's fine, I guess."

"Ma, can I ask you something?"

"Sure you can, sweetheart."

"Well . . . are you and Eddie drinking every night these days?"

My heartbeat throbbed in my temples. I hadn't meant to ask that so abruptly; it had just slipped out. But once I'd said it, I realized that the subject had been quietly brewing.

"Are we drinking . . . ? Well, not *every* night, no. We do have a drink most nights. I suppose Eddie has a little more than me. Why?"

"Well, I mean . . . I was wondering if you had noticed that when you call me in the evening, sometimes you can't remember our conversation the next day. And . . . sometimes you slur your words, especially at night."

I could hear Ma breathing, but she didn't speak. My face felt hot. I couldn't believe I was confronting her about this now, when I needed her so much. What if she took it all wrong? But as I had spoken, I had also realized how true it was, how much this had been bothering me.

"I know you like your drinks, but . . . this forgetting the next day . . . I was just wondering if maybe you should think about cutting down a little bit."

There was another long pause. This was not the first time I had suggested cutting back, just the first time I'd done it as an adult. There had been a rough period during junior high when I'd begged Ma not to drink so much after dinner—she'd kept "falling asleep" at the table. I took a breath and slowly let it out, raising my chin above the receiver, my eyes tracking a passing snowplow as I waited. Finally in a small voice she asked, "Really?"

"Yeah. I've just been . . . concerned, because it's happened a few times now."

I closed my eyes and waited until very quietly, she said, "Well, I guess I'll have to think about that." She sounded so small.

"Ma, I just want to be able to talk to you . . . you know?"

"I want to talk to you, too, of course I do." She paused again. "I will think about it. I'm not saying I'll quit drinking, but if it's true that I do that, I'm glad you said something."

"Thank you, Ma. I love you."

"I love you, too, darling."

We didn't talk any more about nursing or childbirth or any of the other subjects that were on my mind. We hung up awkwardly, and for the rest of the day I felt on the verge of tears.

That night, agitated by heartburn and a sore throat, I slept poorly. As I lay on my left side with my arms and legs wrapped around a long pillow for support, the baby gave a kick. She always seemed to wake up just as I fell off to sleep. And no wonder she was tossing around in there; so much was happening inside her. By now, even before she was able to breathe air, millions of fresh brain cells had migrated into place inside her wet little head, and the sculpting process had begun. As she kicked around in my womb, hearing the gurgles of my digestive system and feeling the sway of my round body, groups of neurons were competing with one another to recruit other neurons into expanding circuits with specific functions: vision, digestion, language, and so much more.

The most active pathways in her brain would be strengthened, and those that lay dormant would atrophy, leading to a great dying off of unused cells. The portions left intact would begin to define our daughter as a person. From the meals I ate to the songs I hummed, everything I did was already leading to changes in the mind of this little pre-person. Right there inside me, she was being tuned by every stimulus I exposed her to, and even after she left my womb and became a separate being, for the next eighteen years I would still be one of the strongest influences in her little life. When I finally slept, I dreamed I gave birth to a kitten instead of a baby girl. I sighed with relief; at least I knew how to take care of a kitten.

CHAPTER 4

In Utero in the Lab

The following Friday, I strode heavily to work. It had snowed most of the night, clearing around dawn, and the thinnest drift of cigarette smoke lingered on the cold air, in the wake of the plow that had come through moments before. I followed the frontage road that ran between the old brick Waltham Watch factory and the frozen Charles River, and a blast of cold air buffeted my huge belly as I stepped onto the slippery narrow sidewalk of the iron bridge that crossed the river. Solid ice held the broken reeds and old pilings at the banks, but at the center the ice had thinned, and dark water moved below the crust.

I felt a sudden pressure against my belly from the inside, and it slowly moved all the way from one side to the other. I paused, reaching deep into both pockets of my coat to cup my gloved hands under my belly. I breathed in and slowly exhaled a puff of vapor into the cold air. What does a contraction feel like? I wondered. Would it feel like a cramp? A squeeze? Sharp pain or dull ache? But it was too early for that, and my little companion was still again. I longed to return to our scruffy one-bedroom apartment and crawl back under the down comforter with Patrick, who would still be nestled there for at least another hour. I sighed and whispered to the little pre-person in my belly: "Well, come on, then, let's get to the lab."

I had spent most of my pregnancy in dark, crowded laboratory rooms full of glowing, blinking electronic recording devices, studying how nerves connect to the heart. For the better part of each day, I sat counting contractions—not my own, but the contractions of heart cells from newborn rats, cultured in a petri dish with neurons. Single heart cells will "beat," or contract, all by themselves just sitting in a dish of salt water with a few additives, but the heart of a mature mammal is controlled by nerves that can make it beat faster or slower. To study how the connections formed, I dissected hearts from newborn rats, softened them with an enzyme to make a slurry of cells, and transferred a tiny squirt of these into culture dishes with similarly treated live neurons. This allowed me to observe and manipulate single neurons as they contacted single heart cells and came to control them. I was studying how these nerves grew in, connected to the heart, and came to control its beat rate as the young rats matured.

Science hadn't always been so important to me that I would get up at six a.m. to spend my day in a room without windows. I came to my interest in neurobiology cautiously, circuitously. I was one of those students who said things like "I'm no good at math and science." I had studied psychology in college mostly because the coursework was easy, so I could focus on more important aspects of life, like the social dynamics at my student co-op. But as early as my freshman year in high school, I had already begun to wonder how the brain worked.

Like most people, I had accepted that the brain was the seat of the mind, but when my science teacher described this three-pound soggy loaf buffered by liquid and encased in a hard cranium, I wondered how it could possibly contain anything as complex as a song, as rigid as a theorem, as ephemeral as a dream. How did the brain create a thought and hold it as a memory, year after year?

Intuition told me the mind was a physically integrated part of the body, rather than the separate entity people referred to, as in constructs like "What's on your mind?" and "It's all in your head." Maybe this was partly because I was brought up with no religious framework; I had no reason to believe that I would ever have a thought, a memory, or any form of existence outside the physical plane. But beyond that, my own thoughts were simply inseparable from my heart, my bladder, my breath and pulse. This didn't make the mystery of the mind any less fascinating for me. It just convinced me that understanding was within reach, if I could only grasp a bit more biology.

My high school biology teacher taught me that the typical nerve cell, or neuron, is like a bulb with a crown of tiny, multiply branching cable-like fibers at one end and one long, exceedingly slender one at the other. The bulbous center is the cell body, or *soma* (Greek for "body"), and contains the machinery required to maintain the cell in working order. The fibers are for communication between neurons. The branching fibers at one end (dendrites) receive messages, and the one long one at the other (the axon) passes them along. In this way, information is passed from neuron to neuron throughout the brain and body.

When two neurons meet so that the axon of one may pass information to the dendrites of the next, the connection they form is called a synapse. Synapses are everywhere. When neurons at the tip of a finger receive the message "too hot," they automatically relay that information via a synapse to another neuron that signals muscles to flex, immediately pulling the hand away from the flame. When we stub a toe, we perceive pain because neurons in the toe communicate the incident to neurons in the brain via synapses. Simple reflexes can involve as few as two neurons with one synapse between them, but more complex behaviors require thousands or even hundreds of thousands of connected cells. To recall a childhood memory, for example, we rely on a complex collection of synaptic connections working in concert to evoke sensation, emotion, event, and context. To give an idea of how complex the connections between brain cells can get, a single neuron can synapse onto up to one thousand cells and receive input from as many as one hundred thousand others.

As I learned more, I began to see that the nervous system was responsible for the things I valued most in life: sensation, communication, and connection. Neurons form relationships; they join different regions of the body to one another, and they attach the inner world of the body to the world outside. Neurons and their synaptic connections bring the world to us in Technicolor, allowing us to receive, process, and transmit information, coordinating our muscles and our minds so that we may walk, talk, think, kiss, sigh, and remember. What could be more gratifying than understanding how we connect, both internally and externally, to our world?

The phenomena I studied had never been more relevant to my life than they were the year I got pregnant. I had been interested in the way neurons grew and connected to one another for a decade, but that year my research had begun to engage my interest in a new way, with an increased intensity; the real processes I was attempting to simulate in my reductionist experiments were taking place in all their complexity moment by moment inside my baby's developing body. Neurons originating near her spinal column extended axons out to form synapses all over her body, including her heart. Bundled by the hundreds, these axons formed the nerves that signaled her heart to beat faster when it needed quick energy and more slowly when it was at rest. These synapses were forming in the petri dishes in my laboratory incubator, and also throughout my daughter's nervous system, inside of me.

As I entered the lab that morning, the first thing I did after flipping on the lights was to turn on the water bath, a small electrically heated tub that I used to gently heat several translucent plastic tubes of pink growth medium to body temperature. The nutrient liquid that bathed my cell cultures had to be refreshed regularly, and it was important not to shock the living cells with refrigerator-cold medium when I "fed" them.

As I waited for the medium to heat up, even though I'd arrived before eight a.m., I was anxious. With a brand-new being taking shape inside me, biological processes had never felt more relevant or exciting—and yet my research had suffered because of it.

Things had changed drastically over the last several months, and increasingly I felt I had been falling short at work. One day I might feel exposed, vulnerable, and the vibrant day could startle me raw. The next, my elephant of a body would engulf my mind, coating it in soft warmth, sleepy hormonal goo, so the world became muted and dreamy. Counting neonatal rat heart contractions on a black-and-white video monitor all day in a stuffy, darkened room was like counting sheep when I was already aching for a nap. More than once I had set out to do an experiment, only to wake up an hour later with my head resting on the black slate lab bench, a small puddle of drool accumulating around my flattened cheek.

I had made a disconcerting discovery early on, something no one had ever warned me about in grad school: pregnancy causes microscope sickness. I hadn't suffered much from morning sickness, but in the first months of pregnancy, looking through the microscope had become the visual equivalent of sitting in the backseat of a Chevy Impala for a fast drive down a twisty road. Five minutes into a dissection, the world would lurch forward, the floor would drop out from under me, and I would have to step outside for air before my breakfast came back up.

At around six months pregnant, that queasiness during dissections had passed, but now, at eight, I had developed a new problem: it was the mama rats and their little pups. I had always been compassionate with my animal subjects, from the sea slugs of my graduate days to these red-blooded rodents—but I had also taken a pragmatic approach to animal use. I firmly believed that the research was worth the sacrifice. I was a carnivore, I had told myself, so why be a hypocrite? Besides, I didn't believe the mollusks were even sentient, and these mother rats were despicable creatures that always ate half their litter anyway. I had kindly anesthetized the animals and deftly severed their body parts, imagining myself like the practiced butcher in an old-world market or a farmer on the prairie, whacking the heads off of chickens. But the previous week, though my beliefs had not changed, on my "prep" morning I had found myself physically and psychically unable to remove the mother's newborns from her cage.

I had never liked the animal storage facility—the smell of damp wood shavings and the thought of the caged monkeys always made me uneasy—but I entered the basement suite fully prepared, holding a bucket for the baby rats, the basement key, and a pair of gloves. I thought I was ready to shoo the mama rat out of the way and scoop her jumble of little blind pups into my rubber bucket as usual, but I was not. As I lifted the lid to the cage, the white mama rat skittered out and stared at me with her shiny red eyes, trembling as she guarded the huddle of writhing pink bodies in the corner. Then she just stood there, twitching her narrow snout at me, and it was as though our minds melded and she took telepathic control of me in an act of fierce trans-species motherly command; for the first time, I saw her as an anxious, protective mom.

I had known that ferocity, that impenetrable defense; it was the hand that had pulled me up out of the pool when I fell in, that had yanked me back out of traffic before I was hit. As I stared at the small, ferocious beast, I saw my mother, strong and red-cheeked, freckles ablaze with the heat of her righteousness, shielding my sister and me from the unclean man who had approached us on the street; Ma, rearing up and conquering her own mild nature in order to fend off harm. I felt her blood in mine, and that same power now rose in me like floodwater, as my focus drew inward, always guarding, in an imperative that overrode every other: to protect my belly and the new life within. The tears welled up suddenly, blurring my vision, and I stepped back, amazed. I could not do it; I could not take the pups from that rat.

After that I had confessed my new shortcoming to my postdoc adviser, Susan. She looked confused and then tired when I told her what the trouble was. Susan was young—only a couple of years older than I—and she'd had two children as a postdoctoral fellow. When she had been pregnant, she had worked the long hours. She'd given talks at prestigious universities resplendently round and tastefully clad in high-end maternity clothes. She'd returned to the lab just two weeks after giving birth, and she had handled the animals throughout. After a moment's thought, she simply offered to have our technician take over my dissections for a while. She didn't protest, but I had no trouble imagining her disappointment in me. I was disappointed in myself. I wanted to be churning out data and publications like a science machine, but I was not. Yes, I was still working very hard every day, but I had not spent a full weekend or even a full evening in the lab for several weeks. I was slowing down. I was a softened, moist-eyed, sleepy, incessantly hungry mama-to-be. I was identifying with the rats, for God's sake.

Once I had refreshed the medium in my cell cultures, I retreated to the back of the lab, to my microscope and my cells. I could hear the lab waking up outside my door: our lab technician's teasing laugh, the voice of a grad student patiently explaining something to her undergrad assistant, the beep of a digital timer. As I opened my notebook and marked the date, I was thinking about the neurons. Through all the trauma of graduate school qualifying exams, grant writing, and rejected publications, I had never stopped loving those cells. They were like children: blameless, beautiful, existential, sometimes annoying, and yet somehow still perfect.

Under the microscope, submerged in pink growth medium and lit up on a disk of thin glass set into the floor of the plastic petri dish, the shiny round bodies of neurons shone, alone or in bunches, like party balloons. Neurites, the tiny fibers of axons and dendrites that nerve cells use to communicate with one another in culture, extended outward from these bodies. They grew fantastically, like Elastigirl's long skinny arms, each minute contortionist cell reaching out many limbs in slow motion along the floor of the dish until, a few days after I had plated them, hundreds of delicate lines crisscrossed one another in an impossibly intricate web at the bottom of their magenta pool. Meanwhile, the amorphous blobs of heart cells—cardiomyocytes—flexed themselves into wrinkled frowns unprovoked, in an amazing autonomous approximation of a heartbeat: *scrunch, scrunch, scrunch.* If I added too many heart cells per dish, they stuck together in clumps, enormous under my lens, all beating synchronously, jiggling and heaving like some alien life-form.

When the neurites in my experiments encountered heart cells or other neurons, their tiny extensions kept right on growing across the dish as if unaffected, but the two types of cells were connecting; they were forming synapses. I knew this because I could perform a cellular wiretap and listen to them talking to each other.

Neurons communicate by zipping messages along their axons in waves of electrical discharge. These waves are called *action potentials*. Shocking a neuron sends action potentials shooting down its axon like electricity through a cable. If the neuron is well connected to a heart cell, then the signal is passed from cell to cell; shocking the neuron causes an action potential in the heart cell also. When shocking the neuron creates enough action potentials in the connected heart cell, it begins to speed up its contractions in the dish: *scrunch-scrunch-scrunch-scrunch!!!*

I filled a tiny-tipped hollow glass needle with a strong salt solution, clamped it into a special holder, and attached it to an amplifier that recorded the electrical current at its tip. This was my electrode, which I would use to record voltage fluctuations in the neuron, and also to deliver shocks to it. The salt solution, with its charged ions, would carry electrical signals from the cell to my equipment and back again. The room was still semidark, and I sat in the glow of the oscilloscope, with its green trace moving across the black screen in a flat line. Then I used a micromanipulator to make precise adjustments to the position of the electrode tip until it was poised just at the surface of a plump neuron cell body, ready to go inside.

Every time I entered the cell this way, I experienced this moment as exquisitely dramatic, an intimate, emotional ordeal encapsulated in a mere speck of time and space. In the next few seconds, either I would pierce the neuron with quiet finesse and be rewarded with a window into the beautiful life of a cell, or I would kill it. If the electrode tip was too blunt or I moved it too far, or if the neuron was not as healthy as it looked, the body of the cell would open at the puncture point and spill its innards out into the growth medium. The cell surface would become pocked and distorted, and within minutes the entire cell would look like a pile of dust. Sometimes, as I watched a cultured neuron collapse and disintegrate, still hanging off the tip of my electrode as its cytoskeleton crumbled and its neurites retracted, I'd get a kind of *Horton Hears a Who!* feeling—Was somebody large and powerful peering in at us at that very moment, studying the complexity of our wee mechanisms? When our earthly superpowers collided and the planet finally went up in a series of mushroom clouds, would some higher form of scientific investigator sigh and wonder, What caused that cell to die in my petri dish?

The tiniest advance of the micromanipulator brought the sharp electrode tip into contact with the cell surface, causing it to dimple slightly. I murmured, "Okay now, little cell, here goes," and taking a breath, I extended my index finger and then very gently tapped on the manipulator. The electrode tip jumped forward mere thousandths of an inch, just far enough to pierce the cell's delicate membrane. Immediately the dimple in the cell's surface disappeared, the slick membrane sealed up around the glass for a clean electrical recording, and the flat line on my oscilloscope screen jumped to action, lighting up with little blips like you see on a heart monitor in the hospital. But the recording wasn't from a heart or even a heart cell—it was a single nerve cell—and those blips were neuronal action potentials, those fleeting voltage spikes that race along axons by the thousands per second, conveying messages that are passed from one neuron to the next throughout the brain and along every nerve of the body as we experience life.

What an exquisite privilege, to peer inside a healthy living cell and watch it perform; on my oscilloscope in those peaks and troughs of voltage I could see the language of the neuron, its message to the world. Without exception, every time I witnessed those bursts of action potentials and the heart cell contracting in response, I was amazed anew. Yet my focus was shifting. That same curiosity and amazement that I had for the workings of the cell I now felt tenfold for the new life inside of me.

Four weeks after we had conceived, with every passing minute, approximately five hundred thousand new neurons had been born into our little embryo's brain. Over the next several weeks, guided by both hardwired cues and interactions with neighboring cells, these neurons had migrated to specific areas and begun to organize themselves, clumping into little islands. At first these brain regions were not connected; they were like cities springing up without any roads between them, but in the first two trimesters, neurons had extended axons, and those had established synapses with their targets: other neurons, organs, and muscles, her heart. For a time her new synapses were being born at the rate of two million per *second*.

I wanted to learn everything there was to know about my baby, as she turned and kicked there in her fluid nest. I wanted to witness the synapses forming, her neurons reaching out to complete the connection so that her brand-new heart would beat just so. What if I could travel inside my own body, into my uterus, through the amniotic sac, past layers of this fresh being's newly formed skin, muscle, and bone, right into the heart of the water-breathing sea creature that would develop into my daughter? I wanted to swim up to her brain, to visit the beautiful pale folds of her tiny cerebral cortex, the furrowed surface of her cerebellum tucked neatly in at the back of her skull, the curve of her hippocampus like a chubby little sea horse, nestled safely under her sweet temple. I wanted to be right there, watching it all happen, with the swish and rush of fluids around us, the pulse of blood, mine and hers, as I roamed inside her, eager to love every cell, every molecule of her being.

By five p.m. the lab had emptied—the technicians had gone back to Boston to begin their weekend, the grad students were off drinking beer and throwing darts in the student lounge, and Susan had left to pick up her kids. At some point before becoming pregnant, I had loved having the lab to myself like this, but now I was eager to leave. I called Patrick and asked him for a ride home.

"You want to go to the North End?" he asked. "I've got a hankering for Italian food."

But I was out of energy. My eyes were itchy, and I felt a tickle at the back of my throat.

"I wish I did, but I'm too tired."

I leaned my head against the cold window as Patrick drove us home through the dark early evening. Once there, I collapsed on the futon by the window in the living room and stared out into the night. A few snowflakes drifted down into the beams of a de-livery truck's headlights as it hissed past on the wet street. Cock-tail hour had already come and gone in California, and I hoped Ma hadn't killed too many brain cells, but I knew better than to call. I would call her in the morning instead. I was still struggling with her apparent lack of interest in my pregnancy. I felt I'd lost my connection with Ma, and I wanted it back.

Pat had dumped his things in the back room, fetched us a cup of tea, and walked back in to light a lamp, eyeing the twenty-pound basketball under my coat.

"May I pet the Belly?" His voice was tender, reverent.

I nodded. He stepped over to the heater vent to warm his hands, then knelt before me. I reached out to touch his hair as he unbuttoned my coat and gently laid one hand on either side of my belly, then snuggled his cheek down onto the middle with a sigh.

"Mmm, Sprout-Tummy."

It was his Homer Simpson voice.

I sighed too, exhausted by the day in the lab, not to mention the construction site that had for months been booming with activity deep within me.

"Did you know that right now the sprout probably already has more neurons in her brain than she'll ever have again?"

"Really?"

"Yup, more than she'll ever need."

Pat raised his head and contemplated the Belly, taking that in.

"But she's still growing."

"Yeah, and she's still forming new synapses all over the place—too many, more than she needs. But it's like her brain is a lump of clay. As she grows, cells and synapses will get carved away, sculpting her." I thought for a moment. "The connections she keeps, and the ones she continues to make and lose, will make her who she is."

Pat took a sip of tea, and then raised his mug into the air. "Well, here's to her synapses," he said. "Because we know good connections can make all the difference in life."

After that he lowered his head back down and began to hum. His voice reverberated all through me, and the baby responded with a gentle thump, comforted, I thought, by the familiar sound of her daddy-to-be.

CHAPTER 5

Hatchling

One and a half weeks after my due date, the mood in our flat was almost festive. Patrick and I both had work we could be doing, but we had entered a state of giddy waiting-paralysis. Pat had started working on his thesis from home; I had finally stopped my work in the lab, although I had a paper to write, and I'd attended my last prenatal aerobics class.

"Rest, reflect, remember who you are, before all hell breaks loose," my friend Laurel had advised me on the phone from Berkeley. Laurel was pregnant with her second child. "I worked until the eleventh hour the first time," she had said, "and believe me, I regretted it." Leah, my best friend in Oakland, backed this up with "Maybe take Pat out on some dates. Things do change once there's a third person in the mix." I sat around the house writing luxuriously long journal entries, and under doctor's instructions to "eat whatever you want," I devoured bowl after bowl of my favorite comfort food, noodles and butter. As often as the heartburn would permit, I took Pat out to our favorite neighborhood restaurants: Little India, Viet Foods, Pandorga's Ecuadorian restaurant, and the relatively upscale Tuscan Grill.

Around noon on the thirteenth day after our shared due date, I called my sister to ask her how the waiting was going in Seattle.

"Oh, you know—I'm sick of waiting, and I'm worried that work will fall apart without me, but I love getting to sleep in. Oh!"

I heard a rustling sound and some loud knocks as Alice fumbled with the phone. When she came back, she said, "Sybil—I think my water just broke! Josh, come look!"

"Oh my gosh, Alice—"

"You know what? I think I've got to go. I'll call you later!" And off went my sister to have her baby.

I hung up and walked over to the mirror, where I stood observing my stupendous belly in profile. "Yup," I said. "It's about that time."

After I called Ma to tell her about Alice, Patrick and I goofed around for the rest of the day. We sat in the window of a sub shop and smiled across the table at each other. An early spring storm had deposited a foot and a half of snow, but the mountains created by the morning plows had immediately commenced to melt; we watched as bewildered pedestrians outside tried to find their way around them and oblivious drivers sped through puddles, spraying everyone with dirty slush. For a while Pat read *Harper's* and I wrote in my journal. Then Patrick looked at me and said, "Dear Diary, I really love noodles and butter . . ."

I smirked at him, then chuckled. This was a mistake, because he loved to make me chuckle, and now I had encouraged him. As I went back to my journal, he began again. "My deepest, most private noodles and butter," he said, and I giggled again in spite of myself. But I looked up and asked, as I often did, "Do you want to hear?"

"Sure."

"It started out as a journal entry, but it turned into a let-ter . . . to the sprouty."

He smiled, and put down his *Harper's*. I cleared my throat.

Oh, Sprout. Never again will I be able to protect you as I can now, with you wriggling around inside me while my big body filters out the noise and the smoke, avoids the mean people and the big cars, and holds you at a perfect 98.6 degrees, even in the harshest of weather. I hope you move safely through the world. I hope you grow up to be a tough enough, smart enough, lucky enough kid to get through all the crazy events that life will bring to you out here. I hope you're a poet inside, struck by the wonders of this amazing planet; that you get out there and ex-perience the big mountains, crowded cities, lonely deserts, and roiling oceans; that you have just enough sadness in your life to remind you of how lucky you are to be happy most of the time. I wish you the best in life. And as long as I am alive and lucid, I will be here to love you and nurture you; I will do my best to be a stable, solid, loving presence in your life.

I looked up at Pat. He smiled and took my hand. "May I write something, too?"

I handed him my journal and my pen. As I watched him, cu-rious what he would write, I wondered whether our child, too, would write with the tip of her tongue thoughtfully protruding from the left side of her mouth. After a few minutes Patrick slid the book back to me.

Dear Sprout,

I can't wait to see you, to sniff your head, to hold you, to read to (and later with) you. I can't wait to teach you how to

drive, and I hope that, as you grow, you find in yourself a strong,
happy, and capable person. I hope that you will become the
person that enjoys being—whatever self that is. It's a bit odd to
think that, one day, you will be as old as I am now. I wonder
who you'll be at that age, and also how different I will be then.
I imagine I will be well on my way towards my lifetime goal of
curmudgeonhood, and that you . . . will be just about anything:
car mechanic, mother, doctor, musician, carpenter, writer,
student, young Republican (strike that), waiter, cook, bookstore
owner . . . I really have no idea. But I do know that I will love
you.

Here's looking for you, kid.

Pat

But you can call me Dad

At two a.m. the next morning we were still waiting, but now
I lay on a hospital bed behind a green curtain with an IV line
running into my forearm, "just in case." On the phone my midwives had suggested that I come in that afternoon and try a little
dose of prostaglandin, to get things started rather than waiting
out the weekend and possibly another snowstorm. The thought
of Alice in labor on the opposite coast had given me courage. I
wished I could call her and Ma both now, to say my labor would
be starting soon, too, but I had waited too long, and now it was
past bedtime on both coasts. It would be so amazing if our babies
were born the same day. How crazy and exciting would that be
for Ma, to have two daughters three thousand miles apart, both
in labor at the same time? I imagined she wouldn't know where
to go first. Alice and I had agreed weeks earlier that we wouldn't
pressure Ma to decide ahead of time; we'd just assume whoever
had the first baby would get the first Grandma visit. I wondered
who that would be as I lay in the dark and the soft background

contractions that had been gently massaging my belly for days began to gather strength.

In the past few weeks the muscle cells lining my womb had consolidated power. They had built a new network of channels that ran from one smooth muscle cell to the next, connecting the salt solution within each muscle fiber to that of its neighbors. The voltage spike that caused one cell to contract now would instantly pass to all the electrically connected others, causing them all to contract together, and the resulting spasms would be powerful enough to expel our baby.

Meanwhile, oxytocin, that little peptide that had stirred me in the middle of the night nine months earlier, was again making mischief, although this was not the same feeling at all. Nine months ago, the pleasure of touch had encouraged me, had spurred me on, but there had been a choice involved. The major source of oxytocin in my blood had been my pituitary, which was in my brain, and therefore other parts of my brain could influence it. I could have cooled down rapidly, if there had been an urgent knock at the door. Late in my pregnancy, however, my uterus itself had begun to produce oxytocin, and also to stock up on the receptors it would bind to, with the result that I had a lot more oxytocin down there, and I was also far more sensitive to it. Fortified by its newly synchronized muscle cells and this powerful chemical command, my womb had begun to act as an independent agent; I was clearly no longer the one in control.

Objectively I knew we had come there to jump-start my labor, but I still had the hazy wee-hours notion that my midwives might just ease off the prostaglandin, since I was in some distress. Only when they did so and the contractions continued did I realize that I had really gone into labor. Somehow, even as we had checked in to the hospital and begun the treatment, it had just felt like another checkup. I had missed that boarding-the-roller-coaster feeling, that *Oh boy, here we go, this is it.* Now this beast that was my uterus kept squeezing and squeezing, harder each time, and clearly it would not be distracted from its mission. "Wait, wait a second," I heard myself say as the tightening around my abdomen began again, but it was too late; the monster had no mercy, and already I was living from moment to moment. It *hurt.*

The specialized neurons that receive information about sensation—heat, cold, stretch, smell, taste, pain—are called sensory receptors. Some sensory receptors can adapt to a stimulus; if you clap several times in a row next to your friend, the strength of the signal sent from his ear to his brain will decrease with each clap, as the receptors inside his ear adapt to the repeated sound. Pain receptors, however, do not adapt; normally, a painful stimulus, no matter how often it is repeated, will register with the same intensity every time. I figured that must be why every contraction seemed to take me by surprise. Between contractions I tried to tap in to my body's natural resources for coping with pain, beginning with the enkephalins and endorphins, my private stash of homemade opiates, which if my body released them now would not only kill the pain but make me feel happy. I was sure these precious peptides could help me, if I could just convince them to flow into my bloodstream during a contraction.

From years of biology courses, I had a rough textbook cartoon in my head representing the pain pathways in my body, and from years of living in California I had a perhaps-naïve belief in the power of visualization. I talked to my pituitary, that magical bean suspended deep in my brain behind my eyes, and to my adrenal glands, like wee berets flung atop my kidneys, willing them to release their stores of enkephalins and endorphins like an infusion of sweet elixir into my bloodstream. As the contractions took me, I saw the pain receptors, a network of minuscule free nerve endings branching like twigs in my skin, my joints, and the fibrous membranes encasing my bones, now twitching with activity; action potentials in a red-hot blaze sped from my heaving abdomen along their axons to a chain of plump nerve bundles alongside my spine, translating the outrageous grind, spasm, and stretch of my body into an urgent alarm signal to be relayed up to my brain: *pain, searing, aching, desperate pain.*

At the junction between the pain fibers and the spinal neurons that relayed the message to my brain lay a secret weapon: inhibitory spinal neurons. For some reason, in my schema these neurons were neon green. If triggered, they would expel a special neurotransmitter chemical called GABA (neon blue, I don't know why) that could silence the relay neurons. If relay neurons could be silenced, the pain signal to my brain would diminish, and I could rest. Unlike some women, who focus on a song, a picture of a flower, or a loved one as they begin a contraction, I concentrated on dampening activity along the pain pathway. In the quiet moments between contractions, I tried to visualize beautiful, soothing blue GABA pouring into those pain relay intersections, cooling things down, and sweet comforting opiates pumping through every capillary on their way to soothe me. As a contraction began, I grabbed Pat's hand, focused on that cartoon in my head, and prayed.

I have been told that many women forget the pain of labor as soon as they have the baby in their arms, but the red-hot ripping, wrenching, and burning of those last minutes of labor have never left me. Maybe by that time my body had already released all its stores of endogenous painkillers in response to the pain of earlier contractions. Maybe the idea of exerting conscious control over pain processes was a bit far-fetched in the first place, for me who had never so much as taken a yoga class. In any case, I think that last stage of labor forced me to commit my first true conscious act of faith. Somewhere along the line I had engaged in a kind of unwritten contract, stating, in effect, *I know and trust myself, and I am willing to take myself anywhere for this one thing.* What could I do but surrender? I stepped forward into the unknown, extending a bridge of belief. I reached, physically and mentally, for the life of another human being, *my* little human being, in a kind of glorious, bloody, screaming welcome party. It was sweat-faith, physical, whole-body faith, and it required me to give up all notion of control. It was good practice for the years of letting go that lay ahead.

When the baby began to emerge, one midwife held up a hand mirror. At first I didn't see much—just a hairy purple patch—but I concentrated on making it larger, and as I pushed, I saw the convex surface of a head, like a little planet, emerging from within me. I gave it all I had, my eyes bulging with the force of it, and I saw that planet grow and grow. But as the contraction subsided and I melted into the bed to breathe, I watched it recede again, and almost disappear.

"No!" I pleaded. "*Don't* go back *in!*"

The nurse smiled knowingly at the midwife, who spoke quickly.

"That's perfectly normal. Babies may act like turtles, but eventually, you can make them come out."

Sure enough, soon after that I put all my muscle and soul into the long, hard, determined final thrusts. Patrick whispered in my ear, gentle and encouraging, while I pushed. Because he was behind me, it was as if his voice were inside my head. I pushed, the midwife pulled, and out came an amazingly dark purple little being, with slate-blue eyes staring angrily left and then right, not breathing a bit.

"Is she okay?" I asked as she was lifted out of my sight, and the midwife said firmly, "Yes, she's fine. She's going to get some oxygen on the table." I looked up at Patrick for reassurance, and his whole face was red, his eyes wide with outlandish, utterly consuming raw emotion, his mouth pulled outward and down a little at the corners in a sort of happy grimace, and tears streaking down his cheeks. I reached up to hold him, and soon I heard suction sounds, a little cough, and finally a small voice that was Zoë's.

And who was Zoë? Zoë, whose name means "life," whose life occupied mine, whose quiet cry I was hearing for the first time. Inside of me she had begun as a tiny presence, a package of life perfectly forming according to unwritten, unyielding instructions, intimately integrated into the lining of my womb. She had been the pea-sized powerhouse that sat silently sapping my energy, triggering the release of hormones that wreaked havoc on my equilibrium. Later, she had become a wispy sea creature, a fluttering minnow, brushing a fin against my inner abdomen, stirring, flipping, telling me secrets without words. As she had grown, the messages had become more decisive: a jab and a kick, or that long, slow movement of head or limb from one side of my belly to the other. She had touched me and only me; she had known the taste of me, she had breathed it. Any character she might have possessed had been an integral part of *me*, because any inkling of her was laden with interpretation, and I was the interpreter. Even at the moment of her birth, it still hadn't really been only about her; it had been about a great groaning upheaval in *my* body; *my* hip bones creaking with the strain as they were pushed far beyond their accustomed span; *my* tissues stretched past any reasonable expectation for stretch.

But she had broken free, I had pushed her free, and now it was finally all about her. Now anyone could interpret; anyone could feel her kick, anyone could touch her. In spite of Patrick's reassuring arms around me, I felt an enormous helplessness unfold in me, the floor dropping out from under me, incomprehensible loss. I *had* to hold her, I had to have her on me, flesh to flesh, that instant. As the midwife brought out the huge burgundy sack that was my placenta and pricked me with needles of lidocaine to stitch up two small tears, someone finally brought Zoë, bathed and pink now but still naked, and set her gently across my chest. Now we huddled together, the *three* of us, oblivious as everyone else worked around us.

As I lay there with Zoë and Patrick, deflated and yet not depleted, I worried again that I would not know how to do this, that the baby wouldn't nurse or I would drop and break her, and tears slid down my cheeks. But I shut my eyes and realized that at the same time I also felt so strong, almost invincible. Evidently, in those hours of reaching for Zoë, reaching for strength and voice and life, I had also been reaching for my new self, for as I had let go of everything I knew in order to bring her forth, like millions of women before me, in an instant I had changed. That firsthand physical experience of faith required to bring her into being had transformed me. Now it was as if someone had changed all the colors, to new ones I'd never seen in my life—not new combinations of those same old colors but actual brand-new primaries that I hadn't even known to exist.

It occurred to me then that my birth and my sister's had also irrevocably transformed my mother this same way; from the moment we had come into the world, Ma had committed herself to the nurturing, surrounding goodness that had personified her ever since. And by joining her in this, I had joined something bigger than myself. It began with the link between Ma and me; her lifelong loving protection had enveloped me. Now my love enveloped Zoë. As though someone were pulling taut a silken strand that ran between me and every mother, the sense of connectedness grew from there. I could not wait for Ma to come and see her new grandbaby. At the same time, I knew she was already there, with me and in me. "Look, Ma," I whispered, touching Zoë's tiny hand, "look what I have."

A little while later, Pat called his mom, and I tried to call Ma, but she was out and I left a message: "We have a baby girl! I'm fine but really tired. I love you. Call us." I did manage to reach Alice. She had ended up having a C-section the previous day, but little Sam was already nursing like a champ.

"Wow, they almost had the same birthday," I said. "I can't believe we have two more family members, just like that."

"I know."

"So, is Ma coming up to help out? You had him first, after all—and a C-section."

"Actually, no. I asked, and for some reason she wants to wait. She's worried about *Eddie*."

Eddie was not in the best of health, and he had recently developed a swelling of the feet that prevented him from walking very far, but he was certainly capable of taking care of himself at least for a few days.

"That's ridiculous. Just keep on her—Eddie will have to manage. You know, I'm going to get her to come out here when Susie gets married, so you'd better get her up there first. She's got to meet these babies!"

"I already can't wait for them to meet each other."

"I know. I can't believe it. Cousins."

Twenty minutes later, when Patrick offered to take Zoë and let me sleep, I looked up at the nurse, unsure, and she nodded. I could barely stand to give Zoë up even for a moment. I had never felt anything as good as the touch of her on my chest, that pawing little blind-eyed bundle against my naked skin. At the same time, I was immensely relieved to be given permission to put the delicate, needy thing down and sink exhausted into the bed. It was a succinct summary of what early motherhood would be for me: I couldn't stand to let go of her, and I couldn't wait to sleep.

For the next two days we inhabited a warm maternity ward nest under the constant care of what felt like a small village; our every need was met. I learned to nurse with the help of a host of earnest, efficient nurses and midwives. Every hour or so someone would stop in to check Zoë's temperature, to deliver a tray of hot food to my bedside, or to pick up the empty tray from the meal before. Patrick held the baby while I slept, made trips to the cafeteria for the newspaper, and generally picked up all the loose ends and bundled them neatly for me. We had one or two visitors, made some phone calls, and composed many e-mails, but for the most part we kept to ourselves, quietly learning to be three.

Between naps and nursing, I waddled around in my PJs feeling swollen and sore, blissed out and bewildered, and refilled my pitcherful of crushed ice, orange juice, and cranberry juice cocktail at the nutrition station. The particular faces and voices on that ward blur together in my memory, but I will always remember the immense comfort of that crushed ice and juice, drunk right from the little pitcher through a wide-bore straw. I wouldn't feel that well cared for again for a very long time.

When a nurse brought me the forms to sign for my release from hospital care, a big part of me didn't want to go. My whole universe had shifted so profoundly, I was like a newborn myself, and after two days in their care, like a newly hatched duckling, I had begun to imprint on the maternity ward staff. I would have followed them anywhere, happy to stay forever in this world of warm faces, warm beds, and warm meals, but now we were to be dumped, abandoned, with this fragile infant in our care, out in the cold harsh world. How could they possibly conceive of such an outrage? They may as well have told me they were about to toss us into the Atlantic.

Patrick carried the car seat with Zoë buckled tightly in and sleeping. I felt fortified in sneakers and thick sweatpants, but it still exhausted me to walk far, and I obsessed about the car seat. What if we didn't buckle it in properly and we crashed? As I hobbled across the quiet lobby, I watched a man and woman stop and turn all the way around in place to follow Zoë with their eyes as Patrick passed.

"Did you see that?" I asked.

"Yeah. Maybe she's really as beautiful as we think she is."

Either that, I thought, or they'd seen the look of terror on my face and realized that the helpless little pink thing in the car seat was being allowed to leave a safe, warm nest before its mama really even knew how to take care of it.

Over the weekend, spring had come, just like that. All the snow was gone, the roads were dry, and the brilliant sunshine blazed through every window of the car as we very slowly drove the mile and a half back to our home. The safe, solid hospital buildings disappeared behind us, and the huge bright complex world, busy with people and cars and strung with roads and telephone wires, hummed and honked and jostled all the way out to the far horizon. As we pulled up in front of our flat, I saw the hulking SUV of our least favorite neighbor, who had once threatened Patrick, and I took a deep breath, glancing back at Zoë's tiny shape, the reassuring rise and fall of the blankets across her chest. I felt the way I had as a teenager, walking alone through San Francisco streets: acutely aware of every potential danger, of the need to stay on the well-lit street and walk near families and couples. Now, though, that feeling was for Zoë; the imperative to keep her safe informed my every move.

CHAPTER 6

Inklings

The day after our return home, I called Ma and asked her to come. When we had talked to her from the hospital, she had congratulated us and asked all about the baby, but she had not mentioned plane flights. Now in a week Patrick would attend his mom's wedding in Connecticut, and I was still bleeding, sore, sleep deprived, and not so great at nursing. I didn't want to be alone.

"What do you think, Ma? Have you started to plan your visit?"

"Oh, honey," she said, "I don't know. I haven't flown anywhere in quite a while. And you know Eddie isn't doing so well."

This took me off guard. When Alice had mentioned Ma's concerns about Eddie, I hadn't taken them too seriously. Of course she'd have to visit both Alice and me; it was just a question of when.

"Ma, I'm sure Daniel can check on Eddie."

Daniel was Daniel Fletcher, my godfather. He and his late wife had known my parents since college, and he remained one of my mom's closest friends. He lived across town, and he and Ma still walked their dogs together every morning. It was Daniel's daughter, Laurel, who had recommended rest and reflection before giving birth, and she had also strongly recommended a visit from Ma. "Get her to do a big shopping," she'd said, "so she can cook you a bunch of dinners. Then freeze them so you have a reserve after she's gone. That's what we did when Micah was born, and it worked great."

Now I heard Ma sigh at the other end of the line. "I don't know," she said.

"I'm just going to be here alone with this little baby, you know? I'm all vulnerable right now. And *you* know how to *do* this." My throat was beginning to close up, squeezing my voice higher. Why did I even have to explain this to her? But I pushed back that thought and barreled on.

"Ma, can you just talk to Eddie about it? It's only a few days. And I can get Patrick to check the flights for you."

Pat was sitting across the room from me when I said this, and I glanced sheepishly over at him. It was a blatant manipulation. Patrick's involvement in anything soothed Ma. It had actually become one of my peeves; she seemed to trust him, the man, more than she did me, the Ph.D. neurobiologist and electrophysiologist, to deal with anything from travel logistics to appliance repairs. I loathed this internalized sexism, though I wasn't beyond making use of it to get my way. But why did I have to push at all? Of *course* she must come.

"Well, all right," she said. *Finally.* "I can see this is important to you, honey. I'll talk to Eddie and call you back. I'm not promising anything. But would Patrick be able to come pick me up at the airport?"

"Of course he would."

* * * *

Ma arrived the Thursday before Susie's Sunday wedding. She would have time to settle in with us before Patrick had to go, and then she'd stay an additional few days after he returned. Pat drove to Logan Airport to pick her up as promised, leaving Zoë and me on the bed nursing. Just after he left the house, Zoë fell asleep on my lap, and I looked around, realizing that for the first time in a week I was essentially alone in the apartment. I gazed out the window at the newly melted river. I heard a neighbor's car door slam and the bark of a dog. Looking down at Zoë's fair face, her little open mouth, I had a sudden urge to write about her. This would be the perfect thing, to sip something warm and write in my journal as she slept. I gently slid her limp body off of me onto the bed, picked up my favorite mug from the bedside table, and tiptoed into the kitchen to make myself a cup of chamomile tea.

I hadn't been in the kitchen since the previous morning. Patrick had waited on me all day, bringing food and drinks to the bed with a bow and a twinkle. He had cooked a delicious chicken stir-fry the night before while I lay in bed with the baby, then cleared away the dishes as Zoë and I watched TV. Now, as I rounded the corner, I was confronted with a catastrophe: mounds of dishes, crusty plates, three-day-old milk cups with white rings marking the evaporation line. I realized that this had been piling up for days: dishes in the sink, on the counter, even spreading onto the kitchen table.

Damn it. I clenched my teeth. I knew Pat and Ma could always do the dishes later, but I had wanted things nice when she arrived. *Why couldn't he think of that?* I felt bleary-eyed and hotheaded, needy and petulant. I just wanted to be taken care of. I looked down at my mug, then back at the dishes. I thought about just walking back out of the room and pretending I had never seen it, but I couldn't stand the mess. Finally, rolling up my sleeves, I grabbed a sponge and flung on the faucet. *Screw it, I'll do them.*

By the time Pat and Ma returned, my nether regions were throbbing from standing at the kitchen sink for so long—I had not stopped bleeding, and now I could feel more was coming—but I had the dishes done. Zoë had woken up hungry before I'd had a chance to put the kettle on, and as she had begun to cry, so had I. I was back in bed nursing her and feeling sorry for myself when they walked through the door. I was sweaty and tired and more than a little unhappy with Patrick, until he plopped the latest *New Yorker* beside me on the bed.

He was trying. "Thank you," I whispered, as I softened in spite of myself. Then I turned to Ma. "So, welcome! I would get up to greet you, but I'm a bit indisposed."

Here was Ma, a little rumpled from the plane, but very much my ma. She wore khaki cargo pants, a red flannel men's work shirt, and comfortable-looking earthy brown walking shoes. Her graying hair was neatly trimmed, probably by Eddie, to just above her shoulders. She just stood there, gaping.

"And this is . . . Zoë? Oh, honey, look at you. Look at that."

I beamed up at her. "It's her. She's almost ready; just a little more on this side . . ." and I switched Zoë to my other breast. "Come in, Ma, come sit. Do you need anything? The bathroom? Food or drink?"

Ma looked a little dazed. She glanced around the room, trying to decide where to sit. I patted a spot on the bed next to me, and she set her purse down on the floor, still looking uncertain. Then she glanced up at Pat. "I guess some water would be nice, thank you. And can you show me to the bathroom?" It occurred to me that finding the bathroom shouldn't be too big a problem in a one-bedroom apartment. Ever so briefly, with a twinge of discomfort, I simply noticed that Ma was a stranger in our house—a ripple in our pool. Pat and I had had such an intimate week.

By the time Ma returned from the bathroom, Zoë had finished

nursing, and I dutifully patted her until I heard a burp, then turned her so Ma could see her rosy face. I waited for Ma to reach for her, to touch her face or her hand.

"It's okay, Ma—you just washed your hands, you can touch her."

"Well, okay, um, how about this?" Ma gingerly held out a finger and stroked Zoë's left leg.

"Do you want to hold her?"

"Oh, I don't know—I'm not sure I know how, really."

"Ma! What do you mean? Of course you know how!" I leaned forward and planted a kiss on Ma's cheek. Then I lifted Zoë up, cupping her head to support her neck, moved next to Ma on the bed, and set Zoë in her arms. "You just have to support her head—remember?" I watched a little nervously as Ma took Zoë into her arms and patted her. It wasn't a smooth, comforting motion, but rather nervous, a little too fast, and my whole body tensed when I saw Zoë's head fall back for a second before Ma's hand came up. I stood a little closer. "See? That's it."

"Ah. Oh, dear. Okay. Babies are certainly small, aren't they? Aren't you just a little baby." I saw Ma soften as she felt Zoë's warmth against her. She began to bounce her and pace the room. "Aren't you just a pretty, wonderful baby, yes, you are." It was exactly the voice she used with her dog, Nita.

The next day I woke up and turned to gaze at Zoë, asleep in the bed between Pat and me, her tiny lips parted and cheeks flushed. It had only been an hour since the last time I'd fed her, and I hoped she would stay still awhile longer; I wanted to just watch her breathe. I put my face close to hers and inhaled deeply. I couldn't get enough of her scent. Something about the warmth of her skin and the sweet, spicy odor of her scalp made me almost giddy. At first it had been mostly masked by hospital products—diapers and wipes and disinfectants—but now, in our home, everything smelled right, and I fell in love with Zoë all over again every time I sniffed her.

As I inhaled, I was putting new brain cells to work. At the time not many neurobiologists would have believed this—the field was still holding fast to the idea that we are born with a set number of neurons that are never replenished—but we now know that new neurons are born to adults. This is particularly well documented in the olfactory bulbs of the brain, which are in charge of taste and smell. Throughout life they receive a modest but steady stream of fresh nerve cells that are born in a region of the brain called the subventricular zone and migrate into the olfactory bulb from there. What is more, in female mammals like me, who are having sex and making babies, the percentage of these new neurons goes way up twice in life: first during pregnancy, and then again when the milk comes in.

I had already begun to believe that mothers have a unique kind of olfactory intelligence. I called it "smell-intel." While pregnant, I had become exquisitely sensitive to smells that signaled potential danger to that delicate little fetus (coffee, gasoline), and now that Zoë was outside of me, my sense of smell felt crucial to staying physically connected with her (the ceremonial sniffing of her scalp). Of course we would need new intelligence to establish that enhanced sensitivity and discrimination. Maybe after some initial confusion as the new cells hooked up with the old, mothers really could acquire increased sniffy brain power.

I wasn't the only one to bond with Zoë and her delicious aroma, though; a moment later, Patrick rolled over in the bed, snuggled up to us, and sniffed her just as avidly as I had, before flinging his arm around me and dozing back off to sleep.

I believe there is just a general yumminess to the smell of a baby's scalp, especially one who's nursing—and maybe there's an adaptive advantage in this, maybe it encourages all the grown-ups in a group of humans to care for all the babies—but it turns out that fathers and mothers also gravitate to the *specific* odor of their own child. In fact, we get to know our child's scent while she is still in utero. The infant's scent is in the amniotic fluid, and it blends in to the mother's scent well before birth. In studies where a mother and father were asked to smell a sample of their infant's amniotic fluid, both parents recognized the scent as attractive and as belonging to their newborn; they could readily distinguish it from another infant's. So toward the end of my pregnancy, as I exuded Zoë's scent, it prepared Patrick and me both for the feeling we had in those moments of lying on the bed in the gray light of morning, inhaling the scent of our baby's skin.

I heard a rustle and the creak of floorboards, and a moment later Ma came tiptoeing in from the living room, where we'd unfolded our futon couch for her the night before. She had originally intended to stay in a hotel up the road, but when she'd arrived, that had suddenly felt too complicated. She'd seemed uncomfortable with the idea of driving our car back and forth, and besides, the point was for her to take care of me, so she should be with me. Crowded or not, this would be better.

"Hello!" she whispered, and she stood before me in her flannel pajamas, uncertain whether she should cross the threshold, even though there was no door between the living room and our bedroom.

"Hi. How did you sleep?" I said, following her eyes to Zoë. "I'm trying not to wake her up. If she wakes up, she'll want more food, and my boobies are sore." I grinned, but Ma didn't quite return my smile.

"Ah. Well, I usually have a cup of coffee, you know. I was wondering whether you have some."

"Sure. The pot is on the kitchen counter, coffee in the

freezer—help yourself. Maybe afterward you can make me some decaf."

She looked confused, so I explained, "I can't have the caffeine right now because of her."

"Oh yes, that makes sense."

"Oh, and filters are on the shelf above the pot."

She stood there a moment.

"And if you want food, there's plenty in there. Just rummage around all you want."

Finally she headed off. I wondered what she wasn't saying; something seemed to be bothering her. I heard her moving things around in the kitchen, a cupboard opening, then the fridge, and I turned back to look at Patrick and Zoë, both still utterly sacked out next to me. Mmm. This was going to be nice. Maybe Ma would bring us breakfast. I settled back into the bed and closed my eyes.

When Zoë began to coo and smack her little lips, I opened my eyes again to a sunlit room. I gathered up my pillows and arranged them comfortably. Nursing, which I felt should have been the most natural talent in the world, was like learning a new sport; I could perform passably after a few days, but I was still awkward and eternally arranging pillows to get the angle of her head just right. When Ma heard me, she hurried in from the kitchen looking agitated.

"Hey, Ma. Did you get your coffee okay?"

"Oh. Well, I—I just couldn't seem to figure out your coffee-maker."

"Figure it out? It's a drip. It works just like yours."

"Well, I suppose I wasn't sure how to turn it on."

"Jeez, Ma, I'm sorry. Did you at least get some breakfast?"

"I—didn't think of that. I always have my coffee first, you know."

When had she become so inflexible? I looked over at Patrick, who was slowly rousing next to me. I knew he'd stayed up too late, because he still hadn't been in bed when I'd nursed at three-thirty. "Can you help her?" He rolled over with a groan, a long lock of hair caught up on the stubble of his chin, and then smiled sleepily up at me and at Zoë. He nuzzled Zoë's head again briefly before lifting himself out of bed. "Okay," he said to Ma, "let's see what the problem seems to be here."

I listened as the two of them made the coffee. "So, you need a filter. Did Sybil tell you where the filters are? That goes in here . . ."

"Oh! I see. It is like mine."

"Water goes here—let's make a half pot—and now all we need is coffee. Did you get out the coffee?"

"Well, no, I hadn't gotten that far. Thank you, Patrick. I'm so glad you're here to help."

I cringed. *She* was supposed to be here to help. And it was *coffee,* for God's sake. But I was relieved Pat was there, too. And Ma would just have to take charge when he left. Or I'd just do things myself. At least I wouldn't be alone.

After nursing, I settled on the couch in the living room, and a few minutes later, Ma came in carrying a cup of steaming coffee, looking much happier. She sat down next to me and patted my arm, holding out a stack of photos. "I was looking at these the other day, and I thought you and Pat might want to have them."

I looked down. A chubby, fair-skinned baby, with cheeks so big and jowly that her eyes almost disappeared when she smiled, sat in the middle of a blanket spread on our old front lawn: me at nine months. A round-bellied toddler in a red swimsuit, I mugged for the camera beside the pool. In the last picture I was a muscular six-year-old with bright blond hair and hundreds of freckles, hanging from the top of a doorjamb with a rascally grin. I was struck by how strong the genes must be on Ma's side of the family. Zoë already had those almond eyes and rosy cheeks, just like Alice and me, just like Ma. All the way down to our chromosomes, bundled together by special proteins and jammed into cell nuclei all over our bodies, the dips and grooves created by shape and charge and order of individual nucleic acids in my DNA molecules matched Zoë's and Ma's in so many ways, and yet differed from them in others. Ma was so clearly inside of me and inside of Zoë.

"Wow, Ma, thank you. These are great."

"I've been looking through the old albums a lot lately. You know, it may sound strange, but I've found myself missing . . . your dad." Daddy had died over ten years earlier. I knew she'd never stopped loving him, but she had been very private about her grief.

"Really, Ma?"

"Yes. Quite a bit, really. I seem to think more and more about . . . things in the past."

I frowned and tilted my head, watching her face. Her light blue eyes with their Nordic eye folds just beginning to sag reminded me so much of her dad, my cowboy Gramps. From just under her chin to the base of her neck, the skin was beginning to hang a little loose these days, just as his had, and deep creases ran downward from the corners of her mouth, which fell into a frown when at rest, just like my grandma Bama's. Ma had always had an open regard, an honest, friendly look that won the trust of strangers. It was still there but a little faded; she looked more tired, less engaged. Again I wondered if everything was okay with Eddie. In my opinion, she had spent too many years living with and nursing men. First she'd supported my father, who'd been mostly blind from glaucoma by the time he had died of cancer, and since then she'd had Eddie, who, after his strokes, had become so easily frustrated, bossy, even belligerent.

Ma saw my look of concern. "Oh, honey, don't you worry, I'm not unhappy now. I've just been enjoying the pictures, that's all."

At that moment Pat, who had been crashing around in the kitchen, reemerged with a platter full of toaster waffles and slightly scorched scrambled eggs. "Have to keep the milk machine going, or the baby gets grouchy," he joked, handing me a plate. The moment passed, and after that we all ate and looked at the pictures while Zoë slept on the bed, and it was cozy in the apartment, with the windows steamed up and the heat on high for the baby. *We can do this,* I thought. *We can manage.*

After the business with the coffeepot, I began to watch Ma more closely. She was anxious to make herself useful, but at the same time she was hesitant to actually do much. When I asked her to heat up my tea, she needed Pat's help with the microwave. She asked for help finding things instead of just rummaging around the way I would in her house, and she became extremely anxious at the idea of taking my car to the store. I puzzled over that. Why would she balk at the idea of grocery shopping? After I had assured her that my car worked exactly like hers, she did finally venture out to shop, but first she had Patrick draw her a map and go over it several times—and the store was only a mile and a half away. *Well,* I thought, *maybe she's just unused to Boston driving.* I could understand that. Then I asked her to cook.

It was around five p.m. the day before Susie's wedding, and the baby had just begun to cry. Patrick had gone out to buy a belt for the wedding; he would be leaving very early the next morning for Connecticut and had to have everything ready. I was tired, a little grouchy, and starting to get hungry for dinner, but I had to nurse. Ma looked up from her book as I rearranged the pillows for the thousandth time and lifted Zoë up.

"Can I get you anything, honey? A glass of water?"

"Well . . ." I felt awkward broaching the topic of dinner. I had hoped she'd just jump in, in her old take-charge way, roll up her sleeves, and put together a dinner for us. I remembered her cooking dinner when I was growing up, the way she whisked food in from the kitchen: a little bit flushed, moving swiftly from frying pan to saucepan to plates and silverware. She always set out fresh vegetables if we got hungry before dinner, and she often got up during the meal to refill a milk cup or replace a fallen fork. Even as a child, I was impressed that she could come home from a full day in the classroom and still muster enough energy to feed us like that. "Well," I said, smiling shyly, "I was wondering how you would feel about cooking tonight."

"Uhh . . . I don't . . . um, I don't know," she stammered.

"We have plenty of food. There's tofu—I feel like I need protein tonight."

Ma looked distraught. Maybe tofu wasn't the thing.

"And I think we still have some chicken breast in the freezer, and plenty of veggies in the crisper."

I waited. As I did, Ma looked away from me, and her face and neck slowly flushed. Then, very deliberately, almost defiantly, she turned back to face me.

"Well, in truth, I don't really cook anymore."

I waited a beat, but she did not elaborate.

"What?" I asked, unable to articulate my amazement. Just then Zoë startled and jerked away from my nipple with a painful *pop*.

"I haven't really cooked in a while now," Ma reported.

"What do you mean? Everybody cooks." The day before, I had mentioned Laurel's idea of having her make us some meals to freeze, and she hadn't said anything then. I felt my temper beginning to rise, and I ducked my head, trying to calm myself as I shifted Zoë to the other side. "It's okay," I murmured to Zoë, "there you go, honey. Drink up." Then I slowly raised my eyes to meet Ma's. She took a breath.

"Well, in the past couple of years Eddie has been cooking more. He likes to cook for me. So I haven't been cooking lately. Not since I can remember, actually."

I stared at her. Two years earlier when we had visited, Eddie had insisted on making us a catfish dinner. He served us each a hulking grayish lump of fish on a bed of plain white rice, with a few exceedingly tired green beans on the side. Always happy to be cooked for, I ignored the appearance and tucked in optimistically, taking a large bite of fish. I started out chewing enthusiastically, but as I did, I realized the fish was slightly overripe, drippy, and utterly raw on the inside. Pat and I eyed each other across the table, simultaneously raising our napkins to our mouths, and secreted the fish away in horror. Ever since then we'd had a running joke about that dinner. "You'd better be good," I'd say, "or you'll get Eddie's catfish for dinner." In mock horror he'd cry, "Not the catfish! Please, I'll do anything, just don't give me the catfish!"

Ma, on the other hand, was a confident cook. She didn't get too fancy, but she knew how to run a kitchen. In our house, no one would know where to find a Bible, but Rombauer's *Joy of Cooking* was always at hand. For dinner we had pork chops, roast chicken, fried rice, enchiladas, spaghetti with a hearty meat sauce. She made our birthday cakes from scratch, canned the tomatoes from her garden, and produced pies from the fruit in our backyard all summer long.

Watching her, I'd assumed that one day Alice and I would be just like her, bustling around the kitchen with families to feed, and she'd loved teaching us. As soon as I was big enough to stir, she had me helping with cookies, even though I always ate half the dough before they made it into the oven. She taught me to cut the apples thin for apple pie, to dot the fruit with butter before laying on the top crust, sprinkle it with sugar and cinnamon, and gently poke holes through the dough with a fork so the steam could escape. To this day I can almost feel her warm body behind me, watching as I cut butter into flour with a pastry knife for pie dough, her voice so clear: "You see how the mixture looks now? Like coarse sand or cornmeal? That means we're ready to add the ice water." And handing me a fork: "Quick now, stir fast, so it stays cold."

Ma couldn't stop cooking; that would have been like ceasing to have freckles, or read books, or love children. Cooking was integral to being Ma. I pouted at her indignantly.

"Eddie does *all* the cooking?"

Suddenly, Ma looked stricken. It was as if she hadn't quite realized this herself, until I forced the issue. What was going on here? I couldn't believe this. Had she become such a helpless old lady now? But she was only sixty-six! What sixty-six-year-old can't cook?

This would have been a good moment for an epiphany. Ma couldn't work a microwave or a coffeemaker; she was uneasy picking up the baby and afraid to drive six blocks in my car; she couldn't cook. I wish I could say that at that moment I realized something horrible was happening to her, that I gently took her in my arms and said, "It's okay, I'm here." I might have been able to help. But I am human, with that natural human tendency to resist fear and change. I didn't leave myself room for an epiphany—I just got mad. I blamed Eddie, I blamed her drinking. I gritted my teeth, and I put my foot down. If she had fallen out of practice, she was darn well going to remember how again, because *this* was unacceptable. In my anger, I allowed myself only the faintest inkling that something was wrong with Ma. I was too comfortable with my established view of her as the capable one, the doer; that was the familiar picture, the accepted reality. Amazingly, even though I studied the biology of the nervous system every day, it never occurred to me that there might be something wrong with Ma's brain.

And no wonder. Altering hardwired thought patterns requires work in the form of attention, which many people find uncomfortable. We tend to avoid change, except when it is essential, as when one must suddenly care for an infant twenty-four hours a day. Familiar activities and thought patterns are stored deep in the core of the brain, in an evolutionarily ancient set of structures called the basal ganglia. The basal ganglia maintain time-honored routines. They function by linking to established neuronal circuits, almost the way a computer program calls up subroutines; the code is already written, so the process takes little energy or attention. This was the part of my brain that allowed me to do the dishes while obsessing about my job, or drive to work while engaged in lively conversation with Patrick. I knew who Ma was; that was all hardwired.

Initially, new information, like all the facts I was having to process about babies and nursing, engages a completely different area of the brain, invoking something called working memory. Working memory is a sort of temporary holding area where new ideas or situations are compared to previously stored information about the world. It is located in a more modern area of the brain called the prefrontal cortex. Working memory requires a lot of energy, and the prefrontal cortex can hold only a limited amount of information at a time. Flooded with so much new input, I was getting pretty overwhelmed, but as I nursed day in and day out, transforming a diligent practice into an easy habit, my basal ganglia were slowly acquiring this function, freeing up my prefrontal cortex for further real-time processing.

This was hard work and required a lot of attention, but it was work I had anticipated and even relished; it wouldn't be long before I could nurse and carry on conversation without any trouble at all. Unexpected, and therefore unpleasant, change affects our physiology differently. When monkeys and human beings perceive a difference between expectation and reality, a special center in their brain's orbital frontal cortex, just above and behind the eyeballs, sets off intense bursts of neural activity, much stronger than the response when something is correctly anticipated. These extra bursts of activity can be seen as a kind of alarm, signaling that an error has been detected. Humans have evolved especially strong error-detection skills, and this is a good trait to have; paying attention to sudden changes in the game plan allows a person to fluidly adjust to new situations. However, stressful news, like the fact that the one person you thought could depend on is falling apart, can generate far worse than a mild sense of discomfort. In fact, it can set off one's deepest animal fear.

One possible reason for this response is the orbital frontal cortex's close association with the brain's fear circuitry, which is located in another deep and ancient brain structure called the amygdala. Activation of the amygdala interferes with the higher intellectual function of the prefrontal region. We become more emotional; our animal instincts kick in. This is why that urge to avoid a thought or situation can feel as potent as the need to escape a vicious predator: change is energy intensive, change is risky, change is scary, change is bad. Avoidance isn't a trivial game when this happens; it can feel like life or death. In the case of facing Ma's disease, I didn't even do it consciously; the fear and frustration just bubbled up inside me. I needed her, this wasn't okay, and she'd better not act this way.

"Look," I said, my face turning red, "people don't just stop cooking, Ma. You have to do this. You just have to. When Patrick comes home, you guys can cook together. It's like riding a bike— it will all come right back to you."

It almost worked. That night Patrick and Ma cooked another

chicken and veggie stir-fry. The next night, in Patrick's absence, Ma would reheat the leftovers—on the stovetop, not in the microwave, which still seemed to elude her. And on the third night, before Patrick returned, at my insistence Ma would make me another stir-fry, exactly as before but substituting tofu for chicken. "So this is tofu," she would say, chewing for a long time before she swallowed. "Huh. I'm not sure I really like it, but it's not bad." *See,* I would think doggedly, *she's fine; she just needs to do more. She's gotten rusty.*

Before light on the morning of Susie's wedding, I was up nursing Zoë when Pat kissed me good-bye. "You'll do fine," he said.

"Okay, I know, it's only a day, right? So give your mom a big hug, do your bit, and drive carefully—but hurry right back."

A few hours later, Ma and I began our day together. I had begun bleeding again, and I was worried I'd torn my sutures. At least Pat had made her coffee last thing before he left, so I didn't have to get up and do that. Should I rest and have Ma get me breakfast, or was that too much to ask? Should I wait until she offered? But she didn't know what to offer. I wasn't sure whether she could handle making scrambled eggs and toast, and yet I worried that she would feel bored and useless if I didn't ask her to do anything all day. Would she wonder why I'd asked her here in the first place?

Finally, when I got hungry enough, I asked her to bring a bowl and some raisin bran to my bed.

"Sure, darling. Is there anything else I can get?"

I thought about the eggs and toast, a cup of hot tea.

"Well . . . do you think you could do scrambled eggs?"

"Oh. I—I suppose I could. Let me see, for that I would need a pan. Do you have a certain pan you use for eggs?"

"No." I heard my own voice go hard and tried to amend it. "I mean, I'm sure you can find one that's good. The little nonstick one is good." I felt a flush come to my cheeks. Ma had changed so much; why was she suddenly so passive? I didn't want to have to talk her through this from my bed. Why couldn't she just go in there and find what she needed?

"You know what? Maybe cereal would be good." As soon as I said it, I felt even more uncomfortable. God, I sounded so manipulative. But in truth I was suddenly genuinely conflicted about what I wanted. I wanted eggs, but I could not face what I felt coming; I could not watch Ma struggle with something that should be easy for her. A part of me knew I should pay attention to this, and another part just shut it down.

"Well, are you sure, honey? I can make the eggs, I'm sure I can."

What was wrong with me? Now she sounded like the Little Engine That Could.

"No, you know what? I actually just want cereal. Is that okay? Really."

As she went into the kitchen, I breathed a big sigh of relief: danger averted. In fact, when she came back into the room carrying my bowl, I had a rush of intense appreciation for her willingness, for her mere presence in the room. She loved me so much. Was she finally tapping back in to her motherly instincts in order to care for Zoë and me? She wanted so much to please me and to help, just as I had wanted her to. She just didn't quite seem to know how to do it. *I'm not giving her a chance,* I thought. *She came all this way.*

Ma spent the day reading, and I read, too, when I was not sleeping or nursing. That was good. Sitting together reading was a comfort, because it was customary for us. For lunch, while Zoë napped, I hobbled into the kitchen and pointed out where to find cheese, apples, bread, mustard. Ma got everything out and made us each a plate to take back to the living room, and I heated water for tea.

During our early, reheated dinner, I watched carefully as Ma poured herself one beer and drank it. Paying attention to Ma's drinking was an old habit of mine that was reemerging as I struggled to account for her strange behavior. All this confusion could so easily be due to drinking. But while there was no denying that she enjoyed the beer and seemed much more relaxed after drinking it, she did not have another.

We turned the television on at six for evening news, which I would never have done alone. As the light began to fade, I think we were both looking forward to more TV followed by an early bedtime, but after dinner Zoë began to cry, and she did not stop.

I tried nursing, but she turned away. I looked in her diaper, but it was dry. I carried her around the house, gently bouncing her the way Patrick did, but nothing worked. This was the first time she'd cried for such a long time. I looked at Ma, filled with a mixture of embarrassment and helpless frustration. Here I was, my first night without Pat, and I couldn't handle it. I wanted to be a model mom, sweet and nurturing, *the way Ma had been,* and here I was, totally unable to console my baby. Another part of me wanted to be saved, to be taught by Ma, the teacher, how to do this, but she just sat nearby, mute, looking worried. I was not who I wanted to be in that moment, and Ma was not who I wanted her to be, either.

As the sun set, I moved into the back room and sat in an old wicker rocking chair. Tears streamed down my cheeks as I rocked and rocked, with Zoë steadily wailing. I closed my eyes and prayed for comfort.

I could remember how Ma had sung me to sleep when I was very small. I knew she had sung a whole array of nursery songs, but the only one I remembered the words to was "All the Pretty

Little Horses." I later found out there had been sad verses that she had skipped; she sang only the sweet ones, and I always loved it when she sang. I loved to fall asleep to the sound of her voice, like a promise that she would be there when I woke. Maybe a song would help now.

Very tentatively, I began to sing the verse I remembered to Zoë.

> *Hushabye, don't you cry*
> *Go to sleepy little baby*
> *When you wake, you shall have cake*
> *And all the pretty little horses.*
> *Blacks and bays, dapples and grays*
> *All the pretty little horses.*

Zoë's crying faltered for a moment, but then she picked back up in earnest. I kept singing. A moment later, I heard Ma walk quietly in. She came to rest directly behind me and took a breath. When her voice joined mine, I thought my heart would break. An imperfect arc connected my childhood to that moment, skipping over all the complex history between as we began the simple verse again together. She stood behind me, and I sat rocking. We continued to repeat that same verse over and over as Zoë howled and howled.

CHAPTER 7

Leaving

I returned to the lab when Zoë was two months old. I had maybe a week of fun; being back from the mothering trenches, I felt like a slightly overweight celebrity back from rehab; everyone greeted me, congratulated me, quizzed me about my new life. I had missed my friends and the intellectual companionship of my colleagues, and I enjoyed the attention. Then in week two reality closed in on me. I felt conflicted: I had to leave Zoë for many more hours a day than I wanted to, but even then I could no longer give my work the long hours it required. The days droned on, and I plodded to and from work barely able to stay awake after nursing all night; Patrick and I rarely saw each other awake. And there was the infernal breast pump.

One morning after I'd been back several months, I sat at the microscope with my hand poised over the electrode holder. It was ten a.m., and I had just lowered an electrode onto a neuron. I held my breath, ready to penetrate the cell, when suddenly *beep! beep! beep!* went the timer.

"Damn it!"

I looked down at my shirt, where two wet patches had bloomed through the fabric of my nursing bra and blouse. Already I had put it off over forty minutes, resetting the timer for ten minutes at a time, the same way Pat hit the snooze button on our alarm clock. My breasts were achingly full. I pushed my chair back with an exasperated sigh and checked to make sure fresh saline was flowing over the cells before grabbing my bag.

Mike, a graduate student rotating through the lab, raised an eyebrow as I hurried out the door with the heavy bag slung over my shoulder.

"Coffee break, already?"

I huffed past without responding. Screw him if he didn't understand; I was late for my date with the sucking machine.

The department secretary had asked me not to leave my pump in the room where I used it during the day, since it was an empty lab not *officially* open for use. "We wouldn't want someone to find your . . . things . . . there," she had said.

"Is there anywhere else I can pump?" Susan had generously offered her office, but that had backfired on the first day, when she'd had an epic closed-door meeting with the department chair. I couldn't exactly kick her out of her own office while she was up for her two-year review. I had tried pumping in the rig room, but people had kept interrupting to use the computer in there.

"Well, there's an outlet in the ladies' room, isn't there?" the secretary pointed out.

Not in a private stall, there wasn't. But even if there had been, I thought, bristling, how would she feel if I suggested she prepare *her* children's lunch in the room where we all defecated? This was the biology department; did we not have a clear understanding of basic hygiene? Why had my hip young professors not conceived of a pumping room for working moms? This was my internal rant as I stomped upstairs, fumbling with my keys.

The heavy door swung open, revealing a row of empty lab bays, their black slate countertops pocked and crusty with use, stretching into the darkness. The fluorescent lights flickered on, two, four, six, the length of the suite. The previous lab had relocated, and while the space awaited renovation, it had become a secret wasteland, visited only by me.

I peered left and right as I moved toward the one remaining desk, which sat in an office space opposite the workbenches and had an east-facing window. Several years earlier, as an insecure graduate student, I had worked in this lab when it was bustling with people. I half expected someone or something to jump out at me as I walked past an eerily quiet row of hulking abandoned high-speed centrifuges, crossing two areas of floor that had been patched and sealed with protective shields and warning tape after radiation spills. A familiar, faintly nauseating odor of phenol exuded from the stark rooms, and the wall clock clicked softly as its second hand stuttered in place over the numeral five. Nothing had changed, and everything had changed.

I lifted the pump out of its bag, set it on the desk with a loud *thunk,* plugged it into the wall socket, then went about attaching machine to tubes, tubes to cups, bottles to tubes, and cups to breasts. Finally, sitting heavily in the one scarred swivel chair, I flipped the switch. As the machine hissed and whirred, I stared out the dirty window at distant hills and a snippet of Boston skyline under the weepy gray sky. I was not happy in Massachusetts anymore. What had happened to our plan to move to California, our big adventure, the fun and travel and Patrick getting to know my old friends and family better? What about being closer to Ma?

I still hadn't sorted out what had happened during Ma's visit, but a constant background anxiety nagged at me. I had the sense that Ma, alone or with Eddie, was fading from my view. When she had left after her visit, Patrick had mentioned that she was beginning to remind him of his grandma, who suffered from a form of dementia. I was more inclined to think she was just in a late-midlife slump. It was a hard time for her. Her close friend Eleanor, my godmother, had finally succumbed to lung cancer just weeks after Zoë's birth. Eerily, then, Ma's old tabby cat had followed in suit. And finally, her beloved golden, Nita, the same dog who had romped on Stinson Beach and licked Patrick's face, had also been diagnosed with lung cancer and died. Now even Eddie seemed to be taking a turn for the worse, though thank God at least he didn't have lung cancer. The losses seemed to have permeated and diminished Ma. What she needed, I thought, was hope, new life: a granddaughter to hold. And what I needed was another pair of hands, a little advice, a nearby grandma to take the baby from time to time. We needed each other.

There was a warm prickle as my milk let down, and it sprayed into the bottle in a series of tiny, perfectly timed squirts. My shoulders slumped, and my throat constricted. It should have been Zoë at my breast, not a machine. The first time I had left her at day care, she had craned her little neck to keep visual contact with me all the way to the door and then let out a cry of outrage when I'd disappeared around the corner. I hadn't even known a two-month-old could see that far. Now, at six months, when I set her down next to the other babies, she beamed happily at her little day-care homies and proclaimed, "Ba-ba-ba-ba-da-da-da-da!" I imagined her teacher Erica, cheerful and efficient, feeding her one of the bottles I had pumped. As grateful as I was for Erica and the other teachers, and even though Zoë was clearly pleased to be there, leaving her with them each morning nearly destroyed me.

Every morning I rushed from home to day care to work, and almost as soon as I'd set up at work, I had to pump. I was lucky if I could gather any data at all, and I ached for Zoë all day. Every weekend I told Patrick the same thing: I wasn't sure I could stand being away from her so much. What if I ended up needing to just stay home with her? He said the same thing each time: that would be okay; we would find a way to do it. Maybe I wasn't up to the task of this tenure-track academic grind after all.

The pump went *shhh-ffft, shhh-ffft;* the clock stuttered. One tear slid down my cheek, and then another. On the home front things weren't right, either. Up half of every night nursing Zoë, I stumbled around in a sleep-deprived daze, while Patrick slowly retreated from domestic responsibility, leaving the cooking and cleaning almost entirely up to me.

The previous evening, I had returned home from work and day care only to find the remnants of Patrick's day spread throughout the house. That morning, before I had left to drop Zoë off, I had quickly done a pile of dinner dishes, swept the hallway, bundled the newspapers into stacks, and thrown dirty laundry into the wash. I had looked forward to returning to the tidy, cozy apartment after work, but Patrick had decided to do his thesis work at home. He had strewn research papers about genetic algorithms and artificial intelligence across the floor, piled stacks of computer code and dirty dishes on the end tables, and strung power cables from the good outlet to his laptop computer, which presently occupied the one comfortable nursing chair. Setting down the car seat as a cranky, hungry Zoë began to fuss, I had launched into full tantrum before I'd even said hello.

"Jesus, what happened in here?"

"Oh, wow, you're back already?"

"Uh, yes. Day care closes at five-thirty. Is there any room in here for us?"

"What do you mean?"

"Patrick, look at this place. Look at me. I don't think you even see me at all. How do you think this place got so clean this morning, by magic?"

"Um, thank you?"

"Yes, thank me! I'm up all night nursing, I get up early to get her ready, and I cleaned this morning while you slept in. Then I go to work all day, and I was looking forward to coming home to a nice clean house. And you do *this*? I live with you, for God's sake. *We* live with you. You have *housemates*."

"Okay . . ."

"God, why do I even bother?"

Suddenly, I didn't even feel like Pat's housemate; I didn't feel like his lover, friend, or coparent. I felt like practically overnight I had become a single mother of two.

"And hello to you, too," he had said. "My day was fine, thanks for asking."

As the machine sucked away, I tried to use my ten minutes of solitude to float above our life, look beyond the daily grind, and understand what was happening to us. What did we need to get back on track? But after the last drops of milk dripped into the bottles, I was still no closer to an answer. I wiped my nose, dashed down the stairs, swung past the lunchroom to nestle two bottles labeled "Zoë's lunch 10/25" at the back of the fridge, and power walked back to the rig room to check on my cells, carrying the pump with me.

On good days I had the same sense of incredulity about the cells in my experiments that I had as I watched Zoë in her bassinet: how could anything be so simple and yet so complex, so common and yet so heartbreakingly beautiful? Even knowing that each and every one of us is carrying more than one hundred billion neurons in her head, I found myself bewitched each day by the behavior of these particular ones in my experiments.

Other days the cells frustrated and worried me just as Zoë could. When they no longer responded to the growth hormones I had added to their bath in the dish, when they got contaminated by bacteria so I couldn't run my experiments, when the heart cells got so overgrown they peeled up off the dish in sheets, bringing the network of neurons with them, I cursed my luck. I wondered where I had gone wrong, and I asked myself why they couldn't just behave so I could finish the experiment, gather the data, and go home.

Now, when I returned to the microscope, I didn't recognize anything. When I adjusted the focus, it looked as though someone had sprinkled my dish with jagged shards of glass. Salt crystals. The solution flowing over the cells had run out in my absence; the cells had collapsed and died, desiccated into a layer of crystallized saline. I sighed and slogged to the incubator for another dish.

That afternoon it was Pat's turn to pick up Zoë from day care. After recording from every last dish of that week's cell cultures, I sat at the computer attempting to make sense of the data until my breasts began to leak and my brain started to hurt, around five-forty-five.

When I returned home, Pat was lying on the couch with Zoë asleep spread-eagle on his chest. His eyes were closed, his nose buried in her soft hair. The house was clean.

I cupped my hand around Zoë's warm, heavy head and kissed the pale skin at her temple. Then I kissed Patrick in the same spot.

He opened his eyes. "Hey."

"Hey," I replied, looking into his eyes, searching for I didn't know what. "Looks nice in here."

"I decided to work at school today—you know, to preserve the peace."

"Well, I appreciate that."

"Good."

He patted the couch next to him, but I stood where I was, frowning a little. We needed to talk, and we both knew it.

It was my usual time to nurse Zoë, and right on cue she began to stir. I turned to her, relieved to have a distraction. She was starting to have a distinct character, with habits, moods, and mood swings, just like her mama. When she was hungry, she started out mildly agitated. If not fed speedily, she would build to pathetic, move quickly on to wildly pathetic, and in the uncommon event that we still hadn't fed her, she would eventually transform herself into a ferocious, inconsolable beast.

She began to stretch her arms straight up above her head and smack her lips. Her early thrashing and reflexes that had begun in utero were gradually being brought under her control as all along the length of her twitching, kicking body, long-distance neural connections were still forming. Flat, greasy cells called Schwann cells and oligodendrocytes were still in the process of wrapping themselves around her axons, insulating them so they could pass signals more quickly. As new synapses formed and others withered, her brain circuitry was maturing and elaborating, dendrites branching and axons reaching, in an enormous beautiful tangle of life. Now she could see, grasp her binkie, turn and track our faces, and voice her woes with purpose and authority. Not only was she babbling fluently, but according to developmental psychologists, she was probably already able to distinguish categories of spoken sound that belonged to our native English, as well as to read lips, correctly identifying the mouth shapes that corresponded to familiar vowel sounds.

"Buh! Buh! Buh!" she said now, holding up her right fist in a tight ball, like a little rebel.

"Right on, sister, power to the babies! Buh! Buh! Buh!" I said, and I scooped her up, moving to the comfy chair to nurse.

Part of me wanted to just sit and marvel at her, but I was stalling and I knew it. Once she had begun to drink in earnest, I raised my head and met his eyes.

"This has been really hard."

"I know. When I finish—"

"Exactly. When you finish. It's what we say."

"What do you mean?"

"Everything will change when you finish. But you said six months and it's been two years. I work so hard to keep things nice, and you're barely doing a thing around here. I need you here, another mature adult, now, not just 'when you finish.' And what about our plans? My 'temporary' postdoc is getting longer than some regular ones."

As I spoke, some part of me loosened. This happened sometimes when I was nursing. I found I couldn't be disingenuous or evasive while I nursed; there was something so basic and sincere about it. I suddenly realized that our previous, rather fanciful decision to move, based more on an adventure than on anything else, was no longer the central issue for me. Over time my motive for the move had changed. Now, I had begun to realize, it was much more about Ma. I still didn't understand what was happening to her, but I understood that she was aging, and I wanted to be there for her. I wanted to spend time with her while I still could.

"It's gone on too long," I said.

Pat's jaw tightened. I thought he was mad, but when he finally spoke, he just said, "You're right."

"I'm right?"

"I haven't been helping around the house, but today I tried. And—"

I folded my arms, exhaling loudly.

"Please don't interrupt me."

"I didn't interrupt you."

He stared at me for a moment before going on. "What I was going to say," he said, visibly straining to keep his voice level, "was that yes, we did say we'd go. And it has been much longer than I thought it would. And it's not fair to you."

I swallowed. "Thank you," I said very softly. He looked like he had something more to say, so I waited.

He leaned forward with his arms resting on his knees, his hands clasped together. "So this is what I propose. Whether I finish or not, we still go. No one ever said I couldn't write my dissertation in California."

"Really?" It was such a relief to hear him acknowledge that we needed to move on. He'd been thinking about this, too, he'd been feeling it.

"Scott's doing it," he pointed out. Our friend Scott had taken a job before completing his degree and was currently in Silicon Valley, writing his dissertation evenings and weekends.

My eyes were filling with tears. "When?"

He thought, closing his eyes again for a moment. Then he nodded to himself and looked up at me, meeting my eye. "How about the end of July?"

I stared at Patrick as he sat there smiling up at me, clearly pleased by the look of surprise and relief on my face. Nine more months I could do. Suddenly, there he was again, my lover, my partner, back in view. He wanted to do right. He had just been caught for far too long in that hardest, do-or-die stage of grad school. I should know how hard that was. At least I had had an adviser who had shown up for biweekly meetings.

"So we go no matter what?"

He nodded.

"You know I do love you, Pat."

"That's good, because you're stuck with me."

•   •   •

In May, I received an e-mail message from a friend of a friend:

subject: job opening.
Sybil,

    I thought of you when I saw this. It's a position at UC
Berkeley, running an undergraduate neuroscience lab course. It's
not exactly a high-profile job, but the hours are decent, and the
benefits are great. You'd definitely have more time with Zoë! It
starts in September.

By this time Zoë was a towheaded, rosy-cheeked one-year-old, and she was plowing through the developmental milestones. She had just begun to teeter around on her feet, inspired at a friend's Easter wedding by some friendly dogs who kept getting away when she simply crawled after them. The little plaster of Paris gargoyle above our doorway had prompted her to pronounce her third word after *Dada* and *Mama:* "DA-doe!" she squealed affection-ately, pointing her chubby finger up at the doorway every day as we returned home from day care. Now she was collecting words at about three a week, grabbing and mashing and rolling objects, compulsively investigating anything with moving parts. Every moment with her, it seemed some new miracle unfolded as this inexorable biological program progressed. I wanted to spend every spare second with her. I sent Cal my CV the following morning and spent the rest of the day smiling to myself. At five-thirty that after-noon, as I swung open the lab door to leave, a new grad student doing a rotation through the lab lifted his head from the DNA gel he was examining and casually inquired, "Leaving early?"

"No," I said gently, "I think I'm just leaving."

CHAPTER 8

Fall Back

What made me sure that my return to California was the right thing? It was not the jobs we had lined up, nor the jubilant letters from my best Bay Area friends welcoming me home, nor even the thought of bringing cheer to Ma's dreary life; it was the land. Here we were aboard a 747 with our baby, entering California, *my* California. After our plane crossed over the Sierras, I leaned across Zoë, who was conked out in her car seat, and peered down at those rolling hills. As the plane dipped lower, I could make out dark green clumps of stout, stubborn grandmother oaks hunkered in the brown ravines. Those strong, scruffy trees and dry golden hills had been a constant in my life for as far back as I could remember. I couldn't wait to visit Ma in my hometown, Martinez, and go for a walk in those hills. I wanted to lean back against the strong, stout trunk of a sprawling oak tree, to breathe in the damp cool air under the bay laurels, climb to the top of a golden ridge to see a blue stretch of bay in the distance. I ached for the licorice scent of anise, the sweet sage, the tangy edge of dry cow pie and horse dung. I watched California roll by below until my ears started to pop from the plane's descent. Then I leaned back into my seat and smiled over at Patrick, who was killing imaginary people in a game of Quake on his laptop. We had made the right decision.

When UC Berkeley had called me back for an interview, Patrick had started looking for a job in earnest as well and quickly received several good offers. After talking to the head of one particularly cool San Francisco dot-com start-up, where he was interviewed for the position of "scientist," his face had lit up. "It was a lovefest!" he had said. With his expertise in artificial intelligence and data mining, he would help his new company generate Web site recommendations. His dissertation was not finished, but suddenly that seemed to matter little. We were moving forward, and that felt good.

We found a sweet little 1920s rental in the Berkeley hills just a week before our respective jobs were to begin, in August 1998. Two immense, stately redwoods presided over the corner of our new backyard. Every evening the fog rose up from the bay and crept past houses and through yards to slowly swirl through their branches, and every morning by eleven or so the sun burned the fog away again, filling our rooms with light and warmth. As Patrick, Zoë, and I settled into our new home, I kept marveling at how little I missed our old one. I fell back in love with the bright, warm days followed by cool, foggy evenings, the whispering leaves, and the buttery bay light.

As we took a last breath before beginning work, Zoë was adjusting to her new day care, where she was ramping up to full-time. Almost every day, after picking her up at school, I plopped her in the backpack for a stroll in the hills above our house. At one and a half, she still napped twice a day, and one of her favorite places to fall asleep was on me. We explored the narrow paths and stone stairways that crisscrossed the hillside neighborhoods among redwoods and eucalyptus. At home when she woke up, I'd strap her into her high chair in our cozy kitchen nook while Patrick and I spoke eagerly about our new positions. Pat would be paid what felt like a ridiculous sum of money to do fun work at his hip new start-up. At Berkeley I would start with a quiet semester developing course curriculum and learning the laboratory exercises I would be teaching in a few short months. I had good benefits for the whole family, and my boss had made it clear how family-friendly the department was when I had visited one day with Zoë in the stroller. It was a good time of life. I could imagine a second child, a deposit on a house, a vegetable garden in summer.

One morning as I unpacked our kitchen, keeping one eye on Zoë, who toddled from cardboard box to box, I picked up our newly connected phone and called Ma. Her home in Martinez was about forty-five minutes away, and I hadn't seen much of her since we had arrived. Patrick had to meet with his new boss the next day, and I thought it would be a good time to drive Zoë out for a visit with Gram.

"We can come over as early as you want," I offered. "Patrick's in meetings all day tomorrow, and Zoë gets me up at six every morning."

"Well, actually, Eddie has been sleeping late. Maybe you shouldn't come too early."

Eddie was suffering from an ill-defined but serious condition that caused various odd symptoms: his feet had swollen and begun to ooze from scabby sores between his toes, his lungs wheezed when he walked or talked, and he had developed an insatiable craving for citrus juice. It wasn't clear to me whether he'd had another stroke or something else was going on, but Ma had become extremely protective of him.

"How about late morning, then?" I said. "Maybe I can take you and Eddie out for lunch."

Ma began to object, but I would have none of it. "Don't worry! We'll take care of Eddie. You need to see your granddaughter."

When we arrived the next morning, Ma hurried out into the yard to greet us as I leaned into the back of the car to pull Zoë out of her car seat. I straightened with Zoë in one hand and the diaper bag and my backpack in the other to find Ma standing uncertainly beside me.

"Oh, look at all the trouble you have to go to," she said before even saying hello. "You really didn't have to come all this way. Goodness, what a lot of equipment you have back there."

"Ma! Hi there." I turned and gave Ma a big hug, with Zoë sandwiched between us. Then I stepped back to let her admire the still-sleepy wonder.

"Isn't she big! Did she sleep this whole way?"

Zoë burrowed into my shoulder and began to cry. I bounced her softly, smiling at Ma, expecting to meet and match her knowing smile. This was a good potential Grandma moment; maybe she would distract Zoë with something shiny, or play peekaboo to make her laugh. But Ma was staring at Zoë. "Is she okay? Tell me what she needs. Should I get a blanket?"

I laughed. "No, Ma, she's a baby. She's just mad because I interrupted her nap."

"What?" Ma shouted over Zoë's cries. She stood with her knees slightly bent and her hands out to the sides, as though she'd just leaped from above. Why was she so tense? I felt a prickle of disappointment. Was this going to be like her visit to Massachusetts, or could we just go back to normal here?

"She's fine! Let's go inside. I'll give her a banana."

We walked across the scraggly lawn to the familiar old pink concrete stepping-stones that led to the front door, and for a moment I was a cackling toddler sticky with watermelon juice. I was a ten-year-old ruffian with a bloody stubbed toe, a giggling teenager ducking out of sight as the boy up the street passed by, a confident college kid home for the holidays, and every subtly different version of myself who lay between. It felt as though a hundred people moved with us through the yard, until I stepped onto the brick porch, when I became a mother, bringing my baby home to visit her grandma.

I proceeded into the living room, smiling around at the knotty pine bookshelves my father had built decades before. The yucky shag rugs were gone, and Ma had bought some halfway decent furniture, but the small rooms were still saturated with the dense, musty odor of stale cigarette smoke, and the light that spilled in from the bay windows still filled the rooms exactly the same way it had when I was ten. As I swung the door closed, something clanked behind me, and I noticed that Nita's leashes were still hanging from a hook on the door, though she'd been dead for months.

Eddie emerged from the dining room, shuffling over in his slippers and a thick blue terry cloth robe that went down to his ankles. He smiled benevolently at me through his Coke-bottle glasses. Tufts of unbrushed gray hair stuck out all over his head, and a piece of something—cereal?—was caught in his large mustache. His potbelly protruded noticeably, lifting the front of his robe up off the floor to reveal his blotchy pale and swollen ankles. "Oh! Well! It's you," he said. His voice reminded me of W. C. Fields's, fading off at the end of each phrase. "It's you, it's you at long last. Imagine that."

I gave Eddie a hug. Eddie always evoked an odd mixture of compassion, mirth, and annoyance in me. I had known him my entire life, but I had never been able to tell when he was being genuine and when tongue-in-cheek. It didn't help that he looked a little like a mad scientist, and the language deficit he'd suffered after his strokes had only made things worse.

"How are you, Eddie?" I asked, stepping back.

"Well, I am super, super, superfluous. And this is just the tip of the iceberg with your mother here—yes, yes, yes. Now I will do my congenial routine and . . . dissipate . . ." His voice trailed off again as he began to make his way into the back of the house.

Ever since his first strokes, Eddie had come off like some bizarre foreigner, using unlikely phrases to convey common ideas. Sometimes his substitutions came out sounding almost poetic, and sometimes just silly. "When we walk the dog, we like to take the rabbit along," he'd once told me. "Rabbit?" I had asked. "Yes, yes, you know," he'd said, his face reddening. "The *tennis* rabbit. We hit the *balls* with the *rabbit*."

I wondered what hidden meaning I might have missed amid Eddie's mixed-up word choice. What did he mean by that "iceberg" comment? His offer to "dissipate" had sounded so forlorn. I worried about that for a moment, as he disappeared into the bathroom, but in truth I was relieved to see him go. In recent years my feelings about Eddie had been further complicated by the fact that more than once I'd seen him lash out at Ma when he couldn't find the words.

"*You* know what I mean," he would hiss at her, "so *tell* them!"

"I really don't," she would protest, half embarrassed and half defiant.

"TELL them!" he would roar, and she would shrink to half her normal size.

Eddie's trouble finding the words he needed—called nominal aphasia—had begun after he'd suffered several strokes. The first one, called an intracerebral hemorrhage, had happened after he'd smashed his head into a windshield during the head-on collision that killed his first wife. Blood leaked into his brain, and pressure built up inside his head. Before he knew it, irreplaceable neurons were damaged beyond repair.

Later, Eddie had had several more strokes, which were probably ischemic strokes, caused by a sudden blockage of the arteries leading to his brain. In an ischemic stroke, the brief blood deficit in the brain lowers the levels of oxygen and nutrients coming to nerve cells via the heart and lungs, with the same result as an intracerebral hemorrhage: damaged and dying neurons. At least one of these strokes must have occurred in the language centers on the left side of Eddie's brain. Whenever I talked to him, I had to resist the urge to shudder. What had those strokes felt like? A tingling in his head, a numbness on one side of his face, a headache? When he reached for an expression, did he find a dark cavern where the words should be? Deep down I knew I would turn mean, too, if I couldn't access the words to express my thoughts and feelings.

Meanwhile, on my hip at that very moment, Zoë was learning to do just that. She was sixteen months old, and she had moved rapidly into the world of the verbal. Her word acquisition was as relentless and compulsive as her toddling had been four months earlier, clearly a biological imperative. Scans of her brain at that point would have shown bursts of activity in her left temporal lobe as she began to verbalize her thoughts and desires. A week earlier she had ramped up to two-word sentences: *Bush hair! All gone! Pu-away!* I had been an expert interpreter of her tone of voice, the category of tears she cried, her body language, but there was nothing quite like having your two-foot-tall daughter inquire, *"Mo bubbus?"* as she sat transferring water from cup to cup in her bubble bath.

Oddly enough, Eddie's strokes and Zoë's development both involved a special kind of cell death. During a stroke, the first wave of neurons dies due to oxygen deprivation (necrotic cell death), but there is a second wave of cell death that often causes more extensive brain damage. This second wave occurs in a very distinctive manner: the cells shrink and shrivel, the complexes of proteins and DNA in the nuclei at the center of the cells condense, and the DNA breaks down. Because these cellular changes are so stereotyped, this distinctive form of cell death, as opposed to that caused by trauma to the body tissues, is sometimes called programmed cell death. It is thought to be the response to a chemical signal released by nearby damaged cells, essentially instructing any cell neighbor to commit suicide. (This can be a useful signal when the cells sending it have been damaged by cancer: the death signal stops surrounding cancer cells from growing out of control. In the case of head trauma or ischemic stroke, however, it rather backfires.)

During development, while nerves are first establishing proper connections with all the muscles and organs of the body, active cell death also plays an important role. Hundreds of neurons send out tiny axonal fibers that are bundled together to form the nerves that will stimulate and control target tissues, commanding the heart to beat, toes to curl, thoughts to flow. When the axons reach their targets, they are given a warm welcome: the cells with which they are fated to connect feed them special nourishing proteins that support their survival. Neurons whose axons do not find their proper targets are deprived of these growth-supporting proteins, and they are signaled to undergo programmed cell death; in essence, if they don't make their proper connections, they are instructed to commit suicide. So as the embryonic Zoë had grown from a tiny ball of cells to a hollow sphere and onward, programmed cell death had constantly whittled away at her cells, generating her form—her heart, limbs, digestive tract, nervous system—all of her gradually taking shape until an identifiable Zoë creature came into being.

I headed for my old place at the dining room table, pulling the banana out of my backpack for the little creature as I went.

"Oh, you'd better not sit there," Ma said, guiding me to the other side of the table.

"It's not that you can't," she said when she saw my expression. She lowered her voice. "It's just that Eddie sits there a lot and . . ." She motioned under the table. I looked down and observed a bare, damp area of carpet. "It's his feet. There's nothing he can do about it. It makes a spot on the rug. I've tried to scrub it, but—"

I shuddered. "Ew, Ma, that's awful."

"Well, it is what it is. I just didn't think you'd want to—"

"No."

I was struck by Ma's no-nonsense approach to his icky bodily fluids. She tended Eddie with the same tenderness I expressed when changing Zoë's diaper. I felt a twinge of shame—I should try to be more understanding.

We sat, and I fed Zoë her banana at the table where Ma had fed me as a child.

"Mo nana!" Zoë declared, her mouth already full, and I glanced proudly up at Ma for confirmation of Zoë's brilliance. I felt a little self-conscious, like I was putting on a performance for my mom: *Look at me! I'm a grown up now!* But it wasn't working. Ma wasn't smiling; she didn't even seem to notice that Zoë had stopped crying and spoken. I looked down at the banana for a moment, trying to relax my shoulders. Then I looked up.

"So, how would you like to get out of the house with us?"

"Oh. What were you thinking of?"

"Do you and Eddie want to go out for an early lunch?"

"Oh, you know, I don't think Eddie will want to go anywhere."

"Well then, how about you and I take a walk with Zoë before lunch? I was thinking we could take her for her first little hike in Briones." I watched Ma's face. I thought she would be pleased—Briones Regional Park had been one of her favorite places to walk when I was little.

"You know, I can't go out for too long. Eddie will need me."

"That's okay. We can just go for a little while. You love Briones. We haven't gone in so long."

She looked worried. "Sure—I guess. I don't know if I remember the way, though."

"What? Of course you do!"

Her glasses had slipped partway down her nose, and she tilted her head back to squint at me, her mouth slightly open. She didn't say anything.

"Ma. Of course you know where it is. The little road up past the old Christmas tree farm." She *had* to remember. She had taken my sister and me there all through our childhood and into our teens.

Silence.

"You know, off Alhambra Valley Road, near Reliez Valley, where Barnum and Bailey used to keep the lions and giraffes off-season?"

Ma stared at me for a moment and then gave me a weary, uncertain smile.

I felt a swell of panic, mixed with affection for her. The park had been a favorite family weekend destination for decades. I had to get her out more—she was acting like an old lady. I knew she'd brighten once she got out of the house.

"Come on. We'll be back in time for lunch with Eddie, I promise."

"Oh, I guess so. I do like to walk. But you'll drive, won't you?"

With Zoë buckled into her car seat sucking her binkie and Ma up front next to me, I drove up the steep narrow road past pear orchards, blackberry bushes, and stands of glistening poison oak. At the end of the paved road, I pulled into a gravelly lot and parked. Ma read about leash laws and rattlesnakes at the trailhead while I packed a happily babbling Zoë into the backpack. We stood and perused the bulletin about sudden oak death, the epidemic that had begun to strike down beautiful old trees throughout the state. "DoggoDoggoDoggo!" Zoë cooed, pointing her chubby finger at the picture of a bobcat on the sign. Her new teacher had been reading her *Go Dog Go* at day care.

As we started up the dirt trail, I noticed how, in the patches of sun, I felt almost too warm, while in the shade, I was grateful for my sweatshirt. In a month we'd be resetting the clocks: *spring forward, fall back,* I silently recited. I loved the fall in Briones; it was nowhere near as dramatic as the New England fall, but the air was clear and brisk. Often in September from the top of the highest ridge you could see all the way to Mount Diablo on one side and Mount Tamalpais and the Golden Gate on the other, both looking closer and more brilliant than at any other time of year. I hoped we'd be able to walk up there today.

Ma was so quiet. Was she tired? Upset? I commented on the weather and pointed out features of the park that had changed, but still she didn't speak. I looked over at her. She walked contemplatively, with her head lowered and her hands clasped behind her at the small of her back. She was such a sweet, quiet, heartfelt woman. And she was my kindred spirit when it came to being a California native; we had it in our bones.

The first time Ma had brought Alice and me to Briones, I had been maybe seven or eight. She lured us with the promise of a can of Coke and a Hershey bar if we made it to the top of the hill. We found paths through the woods and across the ridges, ran through fields past grazing cows, found a cool, slippery salamander beside the muddy creek and lovingly passed it to each other. Ma showed us how the cattle had trampled the hills, so that from a distance they looked smooth, but up close they were flattened into a series of level steps. She pointed out the bones of deer in a ravine, cleaned by scavengers and bleached to chalky white in the sun. After lunch the three of us lay silently on our backs watching the turkey buzzards wheel in the sky above us.

Once when I was fifteen, several years and many Briones hikes later, Ma and I went alone. Alice was off traveling with her boyfriend. It was a big day for me; I was carrying a secret, and I smiled smugly to myself as we trekked up the usual uphill trail. We walked in companionable silence through the laurels and buckeyes to an open field where the trail split: ridge one way, canyon the other.

"Which way?" I asked as we crossed into the field.

She turned and looked me up and down. "You made love, didn't you."

My heart jumped. "What?"

"You made love. That's it, isn't it? You and David had sex."

I shook my head, grinning down at the ground, and felt my face go warm. "Yeah, okay, Ma. We did."

"That's wonderful, honey. I'm so happy for you. And everything's okay?"

This was Ma's way. She didn't mince words. She cared. She wasn't perfect, she wasn't always sober, but she saw me.

Now I wanted to see her, too. I wanted her to know what her love had meant to me, and to revel with me in the fact that though I was still her daughter, now we were both mothers.

"You know, I've thought about how I will share this place with Zoë. I can't wait till she's old enough to tromp around up here like we did."

As I said this, my throat closed up. I knew hearing it would mean a lot to her; this land was so important to her, as it was to me. For eight years I had been holed up in laboratories on the

East Coast, homesick, running DNA gels and dissecting sea slugs and newborn rats, and I had missed my mom and this place with a deep ache that never left me. I wanted us to be a team, to revel in the place we had always loved together. But in reply, she just said, "Really. Huh," and trudged on, her head down, staring at the dusty path in front of her.

I felt myself balk. My head throbbed with a wave of disappointment. Had I formed a false ideal of our relationship, idolized or exaggerated it in my years away? Or had something really changed?

"Ma?"

"Yes?"

"Do you miss this place?"

"I guess I don't really think about it."

"Do you . . . have fond memories of how we used to come here?"

"I suppose so."

"Well, I love it here. I'll never forget it."

"Huh."

My cheeks burned. She could have said so much then. How she had missed me. How glad she was to have me back, how good it felt to be here together. I pointed out flowers and trees, the line of the hill broken by dark clumps of oak, a red-tailed hawk high above us. For the next half hour I tried to strike up a conversation again and again, but each time she failed to meet me halfway. I'd had more comprehensive conversations with Zoë. This was like talking to myself.

Eventually, I fell silent. We crunched along the dry path. The baby backpack creaked, Zoë rocked to sleep, and the pacifier dropped out of her open lips. After stooping to retrieve it, I paused, looked up at the silhouettes of oak branches against the pale blue sky, and sighed. This was my place, and I was immensely relieved to return here, but the relief was tinged with longing.

Then, finally, Ma spoke a full sentence:

"Well, I suppose we should get back to see how Eddie is doing, huh?"

We hadn't even crossed the first cattle guard.

As we headed back down the hill, my mind kept turning things over. I was so disappointed. We hadn't really even talked. Well, I had talked—she had nodded. This silence felt all wrong to me. When I thought about it, Ma had not seemed uncomfortable or upset; she'd just seemed not quite aware socially; she'd seemed *blank.*

I thought about the way Ma hadn't shown any desire to touch Zoë or hold her; she didn't seem to know how to *be* with Zoë. It was coming back to me, how hesitant and passive she had been when she'd come to visit in Massachusetts. Somehow I'd convinced myself that she would be more herself back home, but now I saw this was not so. Today I had witnessed Eddie struggling to express his thoughts, and Zoë bursting with her newfound ability to express hers, but what was Ma's struggle? To me she didn't seem to be fighting to find the words; to me this felt more like a paucity of thought itself. Poor Ma. She must need help. I resolved to take her out more, to liven up her life as much as I could.

Over the next three weeks Eddie's condition worsened precipitously. Ma allowed that he was having liver problems. (*No surprise there,* I thought.) He was making less and less sense and sleeping longer each day. To everyone around Ma, it was clear that Eddie was dying, but she just kept plugging away, caring for him through sleepless nights of his wheezing and groaning as though this were a flu that would pass. Finally, though, we got a call one night from my godfather, Daniel, who lived across town and walked his dog every morning with Ma. She had called him a few minutes earlier for help. "She said Eddie couldn't get up to go to the bathroom," he told me, "and his breathing was very bad. She asked if she should just leave him there until morning, but I told her to call an ambulance right away."

I was in the bathtub with Zoë when Pat brought me the phone. He stood there listening in while I talked to Daniel, and before I even had a chance to call Ma, his eyes lit up with excitement and concern. "I'll go," he said without hesitation. "I'll drive out there right now. Your mom needs us."

I looked down at Zoë, who was trying repeatedly to nest a large plastic cup inside a smaller one. A tall stack of bubbles wobbled comically on her small, wet head. *I should go, too,* I thought, *but I'd sure rather stay and tuck Zoë into bed.* Ma trusted Pat; he was her "good man," and he liked playing that role. "Okay," I said, "go," and he quickly stooped down to give us a kiss.

"I live for these things," he told me as he rushed out, and I believed him. When I called Ma to say he was coming, she protested that he didn't have to, but the next morning she would tell me over and over how good it had been to have him there, and what a good fellow I had.

In the hospital over the next few days Eddie proceeded to have several more strokes, his lungs filled with fluid, and finally one afternoon the doctors asked Ma whether to employ extreme measures to keep him alive.

For the weeks leading up to that moment, she had been unable to think. At every turn she had asked me or Daniel for advice: How did one go about obtaining a wheelchair? Did she have to sleep in the same room with Eddie groaning like that? If the doctor said to call when his breathing got worse, how much worse did it have to get? She'd been as passive and befuddled as I had ever seen her, but when it came to that one decision, she surprised me; she was suddenly very clear. No, she said, make him comfortable, and let him go.

I called Alice.

"Eddie's going to die very soon."

"Well, I hate to say it, but—"

"I know, I know," I interrupted. "But it was so sad today." If I saw Eddie as pathetic, Alice was even less forgiving. Like me, she thought Ma would be better off without him, but she was much less subtle and apologetic about it than I.

"Alice, do you feel like Ma has been, I don't know, just kind of passive lately? I mean, with all this Eddie stuff, she's been asking me and Daniel what to do and just following our instructions."

"Yeah, I know what you mean. She's been that way for a while now. But have you seen the way he treats her sometimes? Have you seen how she hangs her head down when he starts?"

"Yeah. I sure won't miss that. You know, I was really struck by the way she finally made this one decision, not to use extreme measures on him, because until then she hadn't made a single choice without consulting Daniel and me."

"Well, at least she got something right."

"Alice!" I was laughing.

"Anyway, I just hope she snaps out of it when he's gone."

"Me, too."

How shameful would it be to just admit that we were looking forward to Eddie's death? I had compassion for him—I did not want to see him suffer, and he was not a horrible person; he did love Ma very much, after all. But in the end Alice and I both believed, or wanted to believe, that when Eddie was gone, Ma's burdens would be lifted. When Eddie was gone, she would for the first time in decades emerge from under the shadow of the men she had loved and served. She could be her own person again. We just wanted her back.

·   ·   ·

Daniel, Patrick, Zoë, and I all joined Ma at the hospital the following afternoon. Eddie lay there huffing on his oxygen bag with his mouth open wide and his eyes shut. He wasn't visibly conscious, but maybe he sensed us—the movement in the room, our voices, the feel of my hand on his forehead. We had to feed Zoë and get her to bed, so we didn't stay long, but we all spoke to Eddie.

"You look pretty sick," I told him when it was my turn. "I hope you get some rest. We're going now, but Ma is gonna stay with you, just like always." Then I told Zoë to say good-bye, and the sound of her high, pure "Bye-bye!" ringing out like a pennywhistle in the gloomy room made everybody turn and smile. I said, "Good-bye Eddie," firmly, with a warmth and finality I wanted him to hear. Patrick whispered, "Good-bye now," and we started out.

Eddie died in the night. The next day Patrick and I both took a personal day from work. We awoke early, took Zoë to day care, and drove out to Martinez to help Ma. The feeling was more bittersweet than outright sad; there was general agreement, even from Ma, that Eddie's death was preferable to his suffering, and hers. Instead of crying or remembering him that day, we made phone calls, bundled personal items into bags, and flushed countless batches of pills of all colors and sizes down the toilet. I laundered the cover for a chair Eddie had soiled, threw out his soggy socks, and folded clean pants into a bag for Goodwill. Patrick called the mortuary, photocopied the will, and helped Ma make a list of people to call. When eventually we drove home, Ma followed in her car. We had invited her to spend as much time as she needed at our house for a while, and she had immediately accepted the offer.

Almost as soon as she entered the house that afternoon, Ma collapsed into a deep sleep on the couch. We had just set the clocks back, so even though it was only a quarter to five, the sun was already low in the sky. Pat had returned to work, and I took the opportunity to walk Zoë in the backpack up Easter Path; at the top there was a small park where I could watch the sun set across the bay.

As I left the house, a swirl of sycamore leaves scuttled across the sidewalk in a brisk little breeze, and I shivered. As I walked, I kept picturing the scene at the hospital the night before. With morbid curiosity, I had peeped into the rooms neighboring Eddie's, and there had been several elderly people there besides him. Some of those others might be wheeled back and forth from nursing home to emergency room to hospital and back again for years before they died. Seeing them all stacked up there like that had brought it home to me that we were all heading in that direction. Just as some magical combination of genetic preprogramming and environmental stimulus drove Zoë to her compulsive acquisition of new words and ever-improving toddler coordination, processes at work in Eddie's body had pushed him through the every bit as natural developmental end-stages of aging.

As Eddie aged, proteins in the connective tissues that made up his skin, his tendons, lungs, and blood vessels had chemically cross-linked, making them stiff rather than pliant. The three layers of his skin had gradually decoupled from one another, causing the outermost layer to sag, and tiny granules of calcium had gathered in the matrix of his cartilage, making it brittle and hard, just like his nails. With disuse, his muscles had atrophied, and connective tissue and fat had moved in to replace the muscle. The cells that had replenished the matrix in his bones for his lifetime had begun to lose the battle against cells that reabsorbed it, and consequently, his skeleton became porous and light; mineral deposits made it brittle, and his weight forced it to compact, so he became shorter and began to hunch.

All these changes were part of the normal aging process. His heart was less efficient, due to its thickened ventricular wall and the added work of pushing blood through partially blocked arteries. His ribs became more rigid, making it harder to inhale deeply, and the surface area of his lungs diminished, so they began to have trouble meeting his body's oxygen demands when he exerted himself. His digestive system stopped making as much saliva, leaving his mouth dry, and stopped absorbing vitamins as efficiently; his kidneys shrank somewhat, putting him at risk for high blood pressure.

Had all these hallmarks of aging made Eddie feel stiff and tired? Did he dislike seeing himself in the mirror? Or had they happened so slowly he'd barely noticed? Although blood flow to his brain was probably not as fast and efficient as it once had been, and neurotransmitter activity in some areas had decreased significantly, these changes would not affect his learning or memory noticeably under nonstressful conditions. His nerve cells had most likely shrunk some, and the complex arbors of branching dendrites thinned, but language impairments aside, Eddie had remained mentally alert almost to the end; he would have been aware of the changes he was going through. They are changes we all go through, as beautiful in their own way as the darkening and thickening of the bark of a tree or the fade and warp of a wooden barn on a winter hillside. They are also as natural and inevitable as the burgeoning and budding of cells in an embryo or a flower about to bloom. His liver was not aging normally; maybe that was what had pushed him over the edge, but maybe not. In the end it didn't matter why he'd passed; he'd simply been on that path, as we all are.

I hope when it's my turn, I die quickly, I thought. *No, I hope it takes me a couple of days, so Zoë has time to find me and say good-bye.* I pictured a grown-up Zoë alone in my hospital room watching old-lady me with my oxygen bag, slowly dying. *But I'll go quickly then, so she won't have to watch me fade away.* It was too horrible imagining her there alone, so I quickly added a sister and brother, on a plane from Chicago; they would arrive the next day, and they would all cry together; they would all laugh and cry.

When I caught myself wallowing there in my imaginary melodrama, I forced myself back into the present. I had reached the little park with the view, and I focused on the bay, which stretched blue and flat to the far shore, where lights twinkled from San Francisco's jumble of slate-gray buildings. To the right, only the tips of the Golden Gate Bridge towers peeped out above a bolus of fog, and farther north Mount Tamalpais squatted, solid black, silhouetted against a deep orange sky. Zoë squirmed around in the backpack to watch two romping German shepherds in the park behind us.

"Doggo, Mama! Doggo run!" she piped, laughing and pointing, and I laughed with her.

She continued to squirm until the dogs loped away down the road with their person, and then as the light faded, all at once, like a little narcoleptic, she slumped against my back, utterly limp, her fists still full of my hair. I took big deep gulps of the clean night air, relishing the weight of Zoë's small warm body against my back as I strode toward home in the gathering darkness.

CHAPTER 9

Vertigo

Alice and Josh and little Sam flew down from Seattle for Eddie's memorial. We all drove to San Francisco, piled onto a Neptune Society boat with Daniel and Eddie's nephew and his family, and scattered the ashes into the bright, foaming Pacific just west of the Golden Gate Bridge.

For three days our house was full of people. Alice, Josh, and Sam took the guest room upstairs, Ma the living room downstairs. With everyone's luggage strewn about and Sam and Zoë romping gleefully through the rooms, the place felt packed and festive, like a vacation rental on a holiday weekend. Alice and I did feel festive; we embraced the opportunity to get the cousins together, cook some good food, regroup the family, and enjoy each other's company. Ma was understandably overwhelmed—I noticed she kept misplacing her book and her toothbrush, and she seemed distracted, frequently losing track of the thread of conversation—but we all rallied around her, coaxing her to read to the kids, taking her for walks in the neighborhood, and generally showering her with love.

After Alice, Josh, and Sam returned to Seattle, the house felt empty and quiet, although Ma stayed at our house for another week and a half, returning home to Martinez only twice for clothes and mail. It felt good to have her there. Every morning as I made the coffee, Zoë padded over to the futon, loudly proclaiming, "Gam wake up!" and stood by as Ma began to stir. It was nice for me to have another morning person to share Zoë with while Pat slept in, and for Zoë, Gram was like a living toy; she was a source of books and entertainment, an intriguing curiosity, a gift.

I watched Ma closely, eager to see if she would reemerge after the long hard months preceding Eddie's death. She had just been removed from a very stressful long-term situation—Eddie had been a difficult person to live with, and their life had become drearier and smaller every year as his health had deteriorated—but I knew there was a good chance she would get better now that it was over. A whole body of research has shown that chronic stress can cause significant damage to the brain. From vervet monkeys bullied by their dominant peers, to baby rodents taken away from their mothers, the brains of chronically stressed animals develop a shortage of the "trophic factor" proteins that support the birth, health, and growth of nerve cells; these animals stop producing new neurons in their hippocampi, and their dendrites become sparse compared to those of littermates living under happier circumstances. Other research supports the idea that one's living situation can and does affect one's brain: marmosets who live in bare, noisy cages have a reduced number of new neurons born and decreased brain complexity compared to those dwelling in large, enriched enclosures that simulate their natural habitat.

The good news is that these negative effects of living under undesirable conditions don't seem to be permanent; the brain cells of marmosets moved to a roomy enclosure full of branches, toys, and hidden food begin to produce more growth factor and make more connections within a month. This intuitively made sense to me. I was looking forward to watching Ma rise up out of the darkness after her release from the Eddie situation. Several months later I would read studies linking sadness, depression, and the loss of a spouse to the likelihood of getting Alzheimer's. One study in particular, where the level of beta-amyloid in the brains of mice made to live in stressful isolation rose to levels almost twice as high as in "happy" rodent brains, would suggest that living alone, if she was stressed, might actually bring on the symptoms of Alzheimer's. For the moment, though, I was still not allowing myself to consider dementia; I wanted Ma back, and I still believed we could have her.

A few days after Eddie's memorial, I did think she was beginning to open up. She began to really take notice of Zoë for the first time; she seemed flattered, if a little befuddled, by Zoë's constant attention. She helped wash the dishes after dinner and asked Pat and me about our work. When we went out to breakfast at Saul's deli on Saturday morning, she put up a good-natured battle with Pat over the check, and she even laughed at our jokes. Feeling optimistic, I thought of a television commercial for skin lotion I'd seen many years earlier: a lady holds a dry brown leaf up to the camera and crushes it into her hand, but miraculously, when she opens her hand again, the leaf unfolds in slow motion, rejuvenated, green, supple, and new. *This is it,* I thought. *This is what she needed, finally. I am sorry for Eddie, but thank God. Now we'll see her make her comeback.*

Pat wasn't so sure. "No, something's still going on with her," he insisted. She reminded him so much of his grandma, he kept saying, "I really think we should take her for a checkup."

Whenever he said that, I got a little mad at him. I couldn't say quite why, since he obviously had her best interests in mind, but I did. "What good would that do? Besides, you haven't even known her that long," I insisted. "Ma's just a quiet person. Sure, she's been a little out of it, but you would be, too, if I had just died, right? She's grieving. And at least she's cutting down on the drinking. I actually think she's been doing better this week."

One day shortly after Ma had begun staying over, she arrived at our house with her usual overnight bag, but she also produced a small blue cooler containing a six-pack of pale ale, which she proceeded to take out and place in our fridge. I wouldn't have been able to predict my own reaction to this simple act, but it was immediate and strong.

"Ma?" I asked. "What's that?"

"I just brought over a little something to drink."

"Well, that makes me uncomfortable."

Did I say that?

"Oh . . . it does?"

Yes, I did.

"Ma . . . I've told you this . . . when you drink, you forget our conversations—you get all blurry. Yes, that makes me feel bad."

"Oh, honey, do you really mean that? It's just a little beer."

"Ma, I am not in charge of your drinking or not drinking. I know it's your own decision. But . . . I guess I just don't want you to do it around me."

She had reluctantly agreed, not to quit drinking but to quit drinking in my presence, in my house. She hadn't brought another bottle into our house that I knew of. And because she was there so frequently, that cut down her intake considerably. I congratulated myself on single-handedly combating her problem drinking. I told myself and Pat that she was on her way to recovery—from Eddie's oppression, from her grief, and from way too many cocktails.

I was so bent on seeing the positive changes that a few days later, when Daniel called, at first it was hard even to hear what he was telling me. He asked me how Ma was doing, and I told him she was really great, all things considered, and I thought she'd return home soon. I asked after his family. He asked about my work. Then there was a pause.

"Sybil, I was calling because I wanted to tell you about something the doctor told me that night in the hospital room."

"Okay . . . nothing bad, I hope?"

"Well, maybe, maybe not. For starters, apparently when Eddie arrived at the hospital, it looked as though he had had a recent stroke."

"Oh, really?"

"Yes. Which I guess your mom could have missed, as he was pretty sick by then."

"Yeah."

"But there was something else, and this is the part I thought you and Patrick should hear. They did some blood tests, and I guess those showed that Eddie was suffering from malnutrition."

"Okay." I thought about the liver problem, the alcohol. Not too surprising.

"The doctor told me it looked as though he had not eaten . . . possibly for as long as two weeks."

I was silent for a beat. Two weeks? "Okay . . . wow. So . . . what was he saying, Daniel? Did he think . . ."

"Well, Ruthie was presumably trying to feed him. At least you'd assume so. But then again . . ."

I flashed on Daniel's description of Ma's middle-of-the-night call, when Eddie hadn't been able to get up. "She asked if she should just leave him there until morning," he had said, "but I told her to call an ambulance right away." Her best idea had been to wait and hope he would get better. In my rush to get from work to day care and then to the hospital that day, I had put that aside, but now I looked at it again. This was Ma, teacher and caretaker, who had cooked our meals and driven our car, who had earned the money and planted the garden. *She hadn't known what to do.*

I had a creeping feeling that I didn't like, a squirmy, hot discomfort. This was the wrong scene, the wrong script. My voice went a little whispery. "She must have at least offered him the food, right?"

"Well, I assumed so, but—"

I could hear the doubt in his voice. On the pad of paper in front of me, I had written "Stroke" and "No food 2 weeks." As we talked, I had retraced the numeral 2 until it had become fat and dark on the page. Daniel had been a voice of reason in my family for my entire life. He had never failed my parents, and I knew I could trust him completely. The hesitation in his voice had been protective, tender. I thought of Pat's more assertive words: "We should take her for a checkup."

"Daniel, do you think something is going on with Ma?"

I listened as he took a deep breath and exhaled. "I don't know. Maybe. You know, I have known your mom a long time. We all have our senior moments at our age—I know I certainly do. And yet—I don't want to scare you, but I have noticed she often loses track of little things, like a conversation the day before or where she put her date book . . . and I just wonder if we might have something more serious on our hands."

I knew where he was heading. Although I had briefly studied the varieties of senile dementia in one grad school seminar, I remembered next to nothing about Alzheimer's disease. Suddenly I felt irresponsible; I should be more knowledgeable about these things. I let the word *Alzheimer's* into my consciousness and allowed it to settle there. But just as quickly, my mind rebelled again. She was far too young for that.

"God, I was just going to tell you how I thought she was doing better."

"Okay. That may be. But this business with him not being fed . . . Look, your mom and I have seen each other through a lot. She's been dealt some big blows—first your dad, and now Eddie, and this could be just depression. But it feels different to me. I am concerned. We might want to have a doctor take a look at her."

"Okay. Okay, that sounds like a good idea."

"I just think we need to keep an eye on our girl."

I thanked Daniel. I made a note to make sure Ma had her yearly checkup lined up for May. I promised myself I would look into Alzheimer's disease and dementia. I had to take charge here. Daniel was gently leading me, as he had all my life, like a second father. *Take responsibility,* his attitude said; *take care of your mom.* I wanted to be a good daughter to her. I wanted to make Ma a priority, to show her and Daniel that it didn't matter how much else I had going on, I was there for her. We'd figure out whether Ma's behavior really warranted more attention. *Probably not,* I told myself, but it was really time to find out.

When I phoned Alice and told her what Daniel had told me about Eddie and his worries for Ma, there was a long pause.

"So she didn't notice anything when he had another stroke?"

"Yeah, although strokes can be pretty subtle . . ."

"But she wasn't feeding him? At all?"

"Uh, it would seem not."

"Damn." She sighed.

"Yeah, and the thing is, at first she seemed to get a lot better, but now I'm not so sure. It's that thing we talked about before—how she's been sort of incapable and passive?"

"Yeah." She sighed again. "Man, I was assuming that was just because Eddie ordered her around all the time. I guess I was really hoping you'd call and say she was doing great all of a sudden."

"I know. Me, too. And I thought she was, at first. But God, Alice, this isn't the Ma I thought I was coming back to. She doesn't hang out and talk to me. She isn't interested in anything. She doesn't respond to me, you know? And she's so anxious all the time."

"I know."

We sat in silence for a moment. I knew Alice hated this as much as I did. We didn't want to look at it. I felt a rush of gratitude for Alice, for the mere sound of her breathing at the other end of the telephone line. *We* didn't want to face it. *We*. Alice understood, she was possibly the only other person alive who really understood the weight of this. Our mom wasn't like this. This was not okay. And Alice felt it just like I did.

Alice and I hadn't exactly been best friends growing up, but there had been key times when she'd been indispensable. As we paused for that moment on the phone, I remembered the feeling I'd had on my first day of kindergarten. My memories of that day were so clear, the way Alice, two years older, had kept close to me all day, solicitous and protective. She'd led me around our little country free school, to the sandbox where some kids were flooding tunnels with a hose, then over to the slide, which was impossibly tall compared to my preschool slide, and to the chicken coops and duck pond; she'd stood behind me proudly, proprietarily, while she introduced me to all the teachers and recruited her friends to watch out for me. I had felt completely safe. As kids, Alice and I had fought like animals almost daily, but when it counted, she was utterly loyal.

That image of the two of us was connected in my mind to another that was slowly forming, of two new children, and I imagined them now: Zoë and a smaller child, holding hands, standing on our new front porch. Ever since I'd had my vision of a grown Zoë with her brother and sister at my deathbed, the idea had been percolating inside me. Now it seemed like the next clear step. We had this nice house, these good jobs, and Zoë would turn two in a few months—the same age Alice had been when I'd come along. I was feeling so grateful to have a sister myself. The idea sank roots and extended vines; the vines wound all through my consciousness. Zoë needed a sibling.

I had been at Cal barely a month, and everything was still new. This was the semester I was to spend developing curriculum and learning the laboratory exercises I would have to teach in a few months. I was teaching myself how to record from crayfish nerves and cockroach legs, how to micro-inject ion channel RNA into frog eggs and culture nerve cells from embryonic chicks. I spent much of each day alone, stopping only to socialize with the others on my floor at lunchtime. The work was flexible, and at first the pace had been relaxed, but I had a lot of material to cover, and master, before the class began in spring. Zoë spent over eight hours a day in day care; every afternoon I rushed away to pick her up at exactly five-thirty, filled with a painful urgency born of missing her all day. And Pat, accustomed to the self-paced rhythm of graduate school, now had a long commute on top of a regular workday. With Ma still staying in the living room, we were all a little stressed.

One afternoon as we came in the door, Ma leaped up and stood anxiously watching me put away the groceries. Every afternoon she seemed bored and fidgety by the time we returned. She also had a new way of hovering next to me and waiting for me to do something that I just couldn't get used to.

"Ma," I said, "can you just relax? Maybe sit down? I just walked in the door, here."

She looked hurt. "I'm sorry, sweetheart. I've just been waiting for you."

"Look, what would you be doing right now if you were home?"

She shrugged, looked at her watch. "Well, I suppose I'd be watching the news."

"Well, put on the news, then! My place is your place—kick back, do what you want."

"Really?"

"Of course. I want you to be comfortable here."

"Well, since you mention it, I think I really would like to watch the news."

She stood there.

"Well?"

"Do you think you could show me how to turn it on?"

One Monday about ten days after Eddie's memorial, Ma decided it was time to return to her own house. Patrick and I breathed a sigh of relief. Two weeks felt like a reasonable amount of adjustment time, and no matter how sweet Ma was, no matter how dedicated we were to taking care of her in her moment of crisis, having a houseguest had been exhausting. We'd had to be quiet earlier in the evening, and we hadn't been able to watch TV because Ma preferred to sleep in the living room. We couldn't walk around half naked or make the bad-taste jokes that we reserved just for each other. We couldn't make out on the couch at night; we couldn't even argue.

She left on a Sunday afternoon, and the next three days were blissfully simple. With just Zoë to cook for and no need to rush home after picking her up, the pressure was off.

"Heyyy, this is not bad!" Pat said Tuesday night as we lay under a blanket on the couch after Zoë had gone to sleep. He was massaging my feet while we half watched a *Simpsons* rerun. "Is this what normal people do after the kids are in bed?"

"Yeah. It's nice to have the couch back, isn't it."

"Couches are important to couch potatoes."

"Mmmmmm," I said. "Let's sleep in on Saturday and just lie around in bed with coffee and the paper like old times."

"Paper?" Pat grinned and ran his hand along my calf. "Who needs the paper?"

"Well, maybe Zoë, for one."

"Oh, that's right, I suppose there is Zoë to consider. Do you think she'd notice if we . . . while she read the paper?"

I smiled at the image of our toddler studiously examining the stock pages while we made it on the bed next to her. "Hey, do you think maybe she'd like us to make her a baby sister or brother?"

I was flirting back, joking; he had started it. But as I said it, I knew I also meant it. And as I spoke, I felt Pat's hand on my leg pause for a second. Then it resumed.

"Practice," he said. "Baby-making practice sounds fine to me."

"You don't want to make another baby with me?" I said it sleepily, easily, with a teasing pout, as if idly joking, but as I did, I felt my stomach muscles tighten a little.

"Well, no, it's not that—are you serious? I mean, we've barely settled in to this house, and your mom has been here constantly."

"I'm sorry about that—"

"No, no, that's not what I mean. But you know, this is just a strange time to bring it up. What if we just enjoy what we have for a while?"

The first time the topic of kids had come up, years earlier, Patrick and I had been in bed surrounded by newspapers, books, and the remains of breakfast. We'd been talking about elementary school: my California farmland hippie free school, and the bullies who beat him up on the school bus in rural Connecticut.

"So, I know you want kids," I'd said, looking down at my hands, "but I guess . . . well, I think we have to talk about when." I was six years older than Pat. I knew that the latest I'd want to get started would be when I was thirty-five, which would mean trying as soon as we were married, when he was twenty-eight.

"I don't want to have kids until I'm thirty, thirty-five," he said. He saw my face darken. "Twenty-eight at the absolute earliest. But I'd really rather wait."

But Pat's thirty-five would have had me pregnant at forty-one. I pressed my lips together for a moment. Then I spoke, a small, quiet voice. It seemed I needed to take up as little room as possible in the room; the charge between us was so enormous. "I don't want to be too old. I decided before we met that I would start trying by the time I turned thirty-four."

He turned away to stare out the window, and I felt a blackness descend around me. I sucked in my breath, massaging one hand in the other, fighting off images of our tortured breakup, the broken engagement, and me alone, waiting in the reception room at the sperm bank. He was silent, and I waited, listening to the leaves rustle outside our window. Finally, he spoke.

"Well . . ." He took both my hands. "I guess that means we start in a year, right? When we're married. Your latest—my earliest."

We had done just that. And look what a beautiful baby we'd made. So why couldn't we start again now?

"Sybil? We don't have to think about this now, do we?"

"No, of course not—it's okay."

We both turned back to the TV. I tried to relax my body. I told myself maybe I'd planted a seed; no need to hurry things, but maybe at least he'd start to think about it. And just like that, it became "It," something I wanted, something I needed to convince him of.

Wednesday afternoon I called to check in on Ma. "I think I'll do some writing," she said. "It's something I've always wanted to do, and now I'll really have the time." I could tell she was enjoying her newfound freedom, but also she seemed to be having trouble filling her days. I made a mental note to buy her a book of writing prompts I had seen in the bookstore.

Thursday night she was really down. "I'm not sure what to do. I walked down to the marina and fed the ducks. I walked with Daniel. But there's just so much *day* in a day these days."

"Oh, Ma." I couldn't stand the thought of that lonely house.

Friday morning the sadness in her voice eclipsed all thoughts of our Saturday morning in bed. "Do you want to come over tonight? We can do pizza again."

It went like that. She got sad, I invited her over, and she never said no. We got pizza Friday, and she spent the night; Saturday morning we went to the deli for breakfast, and she returned home Sunday afternoon. Sometimes we'd take a week off, but then she would get lonely and the next week she'd be back. During that time I saw Ma slipping back into that same funk she had been in before. She was complaining of dizziness. Again she became apathetic and humorless, and sometimes she asked me the same question twice or even three times in one day. And after some weeks something changed between Ma and me. Pat and I had never intended to invite her to spend *every* Friday night with us, but somehow it became her default assumption, and I became the gatekeeper. I still called her every week, just as I had from the East Coast, but now every week there was an underlying tension as we spoke.

"I just called to check in on you, Ma," I would say.

"Well, that's sweet, honey. I'm glad you're calling."

"Have you had a good week? Any news to report?"

"Well, no, I don't think so."

"I had a pretty good week at work."

I'd give her the rundown about the new job, Zoë's developmental achievements, and the weather in Berkeley, but it wasn't easy to engage her in conversation, and eventually we'd reach that point in the conversation that I dreaded, when I ran out of chit-chat and the obvious question had to be dealt with. Was I going to invite her this weekend? (I didn't really want to.) Was she going to ask? (I would feel guilty for not asking first.) I knew she was lonely. She was waiting to be asked, expecting to be asked, and on the rare occasions when Patrick and I had a work obligation or social engagement, I hated telling her; I could always hear the disappointment in her voice. "Oh," she would say, "of course. I understand." And she wasn't trying to make me guilty; she was like a polite child who'd been told there was no more ice cream left after watching all the other children get theirs. You could almost hear her swallow, trying not to cry.

It was intolerable to me to feel pity for Ma, to view her as pathetic. This was my mother, the capable one, the practical one, the doer, who had done so much for me. When I said no, I was wracked with guilt. Somewhere along the line I had absorbed an idea about family that included the image of multiple generations living under one roof—Granny and Grandpa on the top floor, Mama and Papa downstairs with several kids underfoot, and dozens of cousins at the huge kitchen table for Thanksgiving dinner. Who knows where I got the idea—it certainly hadn't been my own childhood experience, or my parents'—but there it was. I could not let go of the image of Ma's place at my table. During the months following Eddie's death, I constantly revisited the idea of having her come live with us, or all of us finding a house to live in together. Even as I had these thoughts, I knew she would never leave her home of thirty-seven years, and also that such an arrangement would pulverize my relationship with Patrick. Already it was taking its toll. And yet . . . I owed something to my mom, didn't I, for taking care of me all those years? Didn't she deserve the same from me and Alice?

My conversations with Patrick had started to have almost the same tone as the ones with Ma, and Ma's anxiety and sadness bled into my own voice as I spoke to him.

"Do we have plans this weekend?" he would ask.

"Not really. But I talked to Ma today, and she's really sad . . ."

"Yeah?"

"She's lonely. I wish she'd hang out with her friends more."

"Well, you should encourage her to do that."

"I know. She doesn't have any plans Friday, though."

Then I would wait. I wanted him to jump in—on either side. I didn't want to fill every moment of our spare time with my mother, but I also couldn't stand to be the one to say no, which felt like abandoning her. Maybe he would insist on us having a weekend together. "Sorry, Ma, Pat insisted," I would tell her. But mostly he deflected my passive-aggressive offers. "Well, so what do you want to do, then?" he'd ask, and I'd eventually say, "I guess we should invite her."

It was around that time that Pat began returning later on Friday evenings, and returning more exhausted, less communicative. His manner of returning changed, too; where before he had come in to give us both a kiss and pick Zoë up right away, now he often went straight upstairs first, "to wash my hands." When I went looking for him a half hour later, he'd be lying on our bed with his laptop glowing in the darkness.

"Why don't you come sit with us?"

"I just need a little downtime."

Then one Friday I found myself in the kitchen at seven
o'clock, so frustrated and overwhelmed, my ears were humming
as I chopped avocado and tossed it into the salad. Patrick was late
for dinner, as he had been every day that week. Here I was, with
hungry Zoë and hovering Ma. I had left work at five sharp, picked
Zoë up at day care, stopped downtown to pick up a special gourmet pizza, fresh greens, and organic ice cream, raced home and
opened the door, only to find Ma rising from the couch, breathless with anxiety. "*There* you are, *finally*!"

She had her own key. She had a book to read. I had to work.
Why could she never understand this, that I was a grown-up with
a job and a child, and I couldn't be home at three-thirty in the
afternoon to greet her? And why couldn't Pat at least show up
before dinner once a week?

I turned from the salad to wash my hands at the sink full of
dishes from the night before. Pat had said he would do them, but
he had forgotten. This was another new pattern, or the resurgence
of an old one. Because my job was close to home and his was
across the bay in San Francisco, I ended up doing all the pickups
and drop-offs at Zoë's day care. I didn't mind that; I wanted to be
close by her. But somehow this task had expanded to include all
the grocery shopping, all the dishwashing, all the vacuuming, all
the gardening. Did he think my job was less taxing than his, just
because he was better paid? I was still working eight hours a day,
just as he was; how could he leave all this to me? And where *was*
he, anyway?

Zoë toddled in from the living room.

"Gam read Grilla!" she said, holding up her well-worn copy
of *Goodnight Gorilla*. I grabbed her and swung her up into my
arms to keep her from touching the hot stove.

"She did? That's so nice, Bunky." I surreptitiously lifted her
shirt and leaned down to give her round belly a quick smooch.

"Does Gram want to read another book? One more?" This was my ploy: get Ma to spend Grandma Time with Zoë while I put dinner together. It had been awkward at first; Ma had seemed more teacherly than grandmotherly. She talked too loud and too formally and didn't joke around or play silly games. The one thing the two of them shared, though, was their love of books.

"Nudder one!" Zoë shouted in her impossibly high voice as I put her down safely out of oven range, and she trotted off to the living room to find Gram, just as Pat's key clicked in the door.

He poked his head in the kitchen and gave me a peck on the cheek, then turned to head upstairs. My cheeks were burning.

"Okay, I guess I'll see you at dinner, then," I said. My voice was too loud and ragged.

Pat stopped on the stairs. "What does *that* mean?"

"It means I guess you don't want to come sit with us, since that's usually the case."

"I just want to wash my hands . . ."

"Yeah, wash your hands, like there's not a sink downstairs." I had lowered my voice, as if speaking to myself. I turned away from him and yanked the stove open to remove the pizza. From the living room, I could hear Ma's voice reciting, "'Brown bear, brown bear, what do you see?'"

"There's pizza," I said, not looking at him, "as usual. There's a nice bottle of sparkling cider, and pizza, and a beautiful salad, and a beautiful little girl, and your beautiful wife all here downstairs waiting for you, if you should ever want to join the family again."

He stared at me. "Okay . . ."

"*Okay.* So go wash your hands."

I took a deep breath and called Ma and Zoë into the kitchen. "Water, Ma?" I asked, trying to keep my voice level as I set Zoë's sippy cup on the table of her high chair.

"Oh yes, thank you. That's great. Isn't this lovely. Thank you, darling." Ma placed her hands in her lap and looked up expectantly, as if waiting for the king to arrive. "Did I hear Patrick come in?"

Patrick did come in. He made pleasant conversation with Ma and gave Zoë lots of attention. He even helped clean up the dishes. But he and I didn't actually speak until after I had helped Ma make up the futon in the living room and Zoë was finally asleep in her little bed. As I took off my clothes, I felt Pat watching me, and I reached for my flannel pajamas. I didn't feel like being seen.

"So," I said as I dropped into the bed next to him.

"Yeah, so."

"So, I'm sorry I was mean. But I don't think I was out of line. It feels like you don't want to be here."

His eyebrows shot up. "What do you mean? Of course I want to be here."

"Well, you used to come home for dinner, for one."

"I was home for dinner."

"Not last night. Not the night before. And tonight you were late . . ."

"What do you mean? We ate dinner together."

"I mean you came in at the last possible second. I'd already put off dinner for an hour."

"Hey! I commute an hour and a half each way. How do you expect me to be here as many hours as you are?"

"I don't. But I expect you to make an effort. If it was me, I would be out that door at five a.m. so I could be there a full day *and* come home while Zoë was still awake. But you can't do that, can you, when you stay up all night shooting people in some video game and then sleep in until eight-thirty."

"So we have different sleep schedules. I'm a night person. Can I help that?"

"Yes, you can help that. You can go to bed earlier. You can make an effort! God, you don't even try."

"That's not fair. Look, I work all day, I bring in three times as much money as you. If it weren't for me, we wouldn't be able to afford this house at all. That's a lot of pressure, having to support you guys. Why don't you give me a break?"

I stared at him. "Oh. My. God."

"What, 'Oh my God'?"

"You actually said that. You actually think your job is more important than mine because you make more money! You think because you happen to get a fat paycheck that what I do is worth nothing? What I do is damned important! For one, I pick up our *child* every night and drop her off every morning. I feed her and shop for us and do every fucking piece of housework there is, and you think just because no one pays me for all that extra stuff on *top* of my *full-time* job—which, yes, is underpaid, but excuse me, who gets the good health benefits for all of us?—you think that I am not as fucking *important* as you?"

"Can you please keep your voice down?"

"God! I just can't believe you can think that way!" My arms were folded across my chest, and I was gritting my teeth so hard I felt my incisors move in my gums.

"Sybil, I'm not judging your work. I respect what you do—"

I took a deep breath and forced it out between my teeth, shaking my head.

"I do! It's part of what drew me to you in the first place. And I *do* want to spend time with you and Zoë. I *do*! I wish *I* could be the one that did all those things with her—but we can't all live off your salary, can we? And look at it from my perspective for a second: I'm all the way in San Francisco, and I'm new at this job, which is really *hard,* and I need to do my best . . . I can't be every-where at once—"

"I'm not saying you—"

"I'm not finished!"

"Fine." I hugged myself, biting my lip hard.

"I come home and I'm tired, and you guys are all jumping at me, and with your mom here we can't really talk, and Zoë starts to cry . . . it just gets to be too much for me, and yes, I do need some downtime at the end of the day."

"Are you finished now?"

"For now, yes."

"Okay. Look, how do you think *I* feel at the end of the day? Do you think I don't need *downtime*? I work all day at *my* hard job, and then I have to pick her up. And cook for her." By this time the tears were streaming down my face. My voice was a harsh whisper. "You're there riding your BART train, reading your novel—that sounds pretty great to me! Think how my day ends. I open the door, and there's Ma, and she's been waiting all day, so I have to help her, and then I have to cook, and then I have to clean, and what downtime do *I* get? Did you think of that?"

There was a long pause.

"No." His voice was small. I was still glaring at him, but his face had softened. I felt my anger fading a little, being replaced by a helpless sadness that was welling up in me. I tried to fight it; being mad seemed preferable, but the helpless feeling began to take over.

"I guess I'm not thinking about your side, either," I said. "Shit."

"I'm sorry, Sybil."

For a second I couldn't say anything, even though I knew it would help. "I'm sorry, too," I finally whispered, my head in my hands.

"It's a tough time right now."

"I know."

He was silent, and I looked up. He was holding his index finger out to me. It was his way of asking permission for the fight to be over. I reached out my index finger and touched the tip of his. He pulled me over to his side of the bed and kissed me, which made more tears come. I lay my head on his chest and let them come, while he patted my back. Everything had gotten so out of control. What had happened to us? I longed for the infatuation and adventure of our early days of dating. We'd been silly then, and busy, too, but not so overwhelmed somehow, and full of admiration for each other.

"You okay?" he asked.

I did a big shaky sigh. "Patrick?"

"Yeah?"

"Do you think Ma will be okay?"

He didn't answer but instead turned and kissed me, very tenderly. I opened to him, seeking relief from my increasing sense of powerlessness. We made love slowly, softly, deeply. I didn't want to use protection, but Patrick insisted. When I came, I cried.

CHAPTER 10

Pivot

A re you sure you have enough gas, there?" Ma asked, leaning over to peek at the fuel gauge as I drove us to the supermarket.

"Sure, there's at least a quarter tank. We can go over fifty miles on that."

"Well, better safe than sorry, you know."

"Ma, we're just going to the store!"

Eddie had been gone a year and a half, and it was yet another hectic Friday afternoon with Ma and Zoë. In our house we now freely acknowledged that Ma was off-kilter, that she had issues, that she was aging. Pat still thought something more serious was going on, but although her regular doctors had agreed that she had some memory problems, they had merely recommended that we "keep an eye on her." I resented the fact that they seemed to treat Ma deferentially, apparently downplaying her memory problems simply because she was an older woman. At the same time, they were providing me with the perfect out, giving me license to look the other way once again; they were the experts, after all.

So I repeated their prognosis to Pat, and we blundered on for months, with Ma in our midst every weekend.

With increasing frequency, Ma's behavior gave me pause. One night she had been talking about Daniel and his late wife, Eleanor, who had been one of her closest friends for over forty-five years and had been gone only three. "Now, what was Daniel's wife's name again?" she'd asked me. She became confused about directions and often lost track of where we had parked the car, although she had no trouble making the forty-minute trip to visit us every weekend. Often she sat for long periods, silent and slack-jawed, staring at the table or at her own feet. And incessantly she worried. If I didn't call every day during the week, she panicked and called me at work, full of angst: "I just didn't know what had become of you!" As it got close to bedtime, she worried that we would not finish dinner in time for her television shows. If I didn't stop the car when she expected me to, she'd suck air through her teeth and go rigid, her brake foot slamming against the passenger-side floor of the car. All these new habits of hers grated—they went against my idea of who she was at her core. Today it was the gasoline.

"Well, whatever you think," she said, shaking her head and looking down at her lap. Then she looked up again. "Look, there's a gas station! Don't you think you'd better stop?"

I glanced in the rearview mirror at Zoë in her car seat. She'd had a long day; a clump of her hair was clotted with something—probably yogurt from her lunch—and the string of pink flowers her friend Maddie had drawn on her cheek with a marker had smeared, making her look like a burn victim. It was only a matter of time before she got hungry or had to pee. Timing was everything with a three-and-a-half-year-old. I looked back at Ma, who was straining against her seat belt to better see the gas station, and I tried to take a calming breath.

"Oh, all right, Ma. But this much gas would easily last me into next week. I don't see why you get so weird about this."

As I swung the car into the station, Ma loudly sighed her relief. "Well, thank goodness. I think you did the right thing."

Now we drove to the Cheese Board bakery for our gourmet pizza and stopped in at the little produce shop next door for greens and an avocado. Friday afternoons at the Cheese Board struck me as Berkeley at its best—folks young and old sat on the grassy median eating goat cheese pizza and drinking Orangina while inside a boisterous circle of Cal professors drank wine and a jazz trio played Cheese Board originals. A young couple clad in Italian leather talked motorcycles with a long-bearded homeless man; kids petted dogs who mooched scraps, and the flower vendor was doing good business. I had the old feeling of longing to do grown-up things, a yearning to have my independence and become an enthusiastic participant in all the fun. Everywhere we went, I kept an eye on not only Zoë but also Ma. She was no longer completely comfortable being asked to stand in line for pizza while Zoë and I went to buy ice cream. I was lucky that they both felt safer close to me and neither tended to wander away, but sometimes I just wished I could leave them sitting on a bench and forge on alone; I wanted to sip wine, to buy flowers, to sit with my friends without a care.

As we got into the car to go home, Ma again leaned over to check the gas gauge. "Oh, good girl," she said. "I'm so glad you keep that tank full. Better safe than sorry, you know."

That night Patrick and I had a small surprise for Ma. Months earlier, when we'd arrived from Massachusetts, she had loaned us ten thousand dollars for the move, and we had finally saved up enough to repay her. After dinner Pat produced the check with a flourish.

"There you go, Ruthie. Repaid in full. With interest."

She took the check from his hand hesitantly, looked at it for a moment, and then looked up again, utterly baffled. "I don't understand. What is this for?"

"To repay our loan. From the move."

"Did I loan you money?"

I looked at Pat, then back at Ma. "Well, yeah, Ma. When we moved. Ten thousand dollars. You can't have forgotten . . ."

But she had. Completely. She fought us on it, she didn't believe us at all. It took us half an hour to convince her to deposit it into her account anyway, and that only because Pat played the Man card, insisting with stern authority until she gave in.

Alzheimer's memory loss typically proceeds in reverse order of acquisition, with short-term memories falling away first and memories of early childhood retained longest. Ma's older conscious memories would remain intact for some time, but new ones were already becoming difficult to hold on to. This is called anterograde amnesia. It happens early on in Alzheimer's disease because the hippocampus and related temporal lobe structures, which are essential to the formation of new, conscious memories, are some of the first structures to be affected by the disease.

Short-term memories—composed of information from the present moment—are extremely transient, stored only for a matter of seconds or minutes. They are thought to consist of temporary shifts in the activity of certain neurons, probably due to changing neurotransmitter levels. We hold on to a bit of extra information only for the short time that these changes in the strength of synaptic signaling are in effect. Like footprints across a lawn, the traces of these memories disappear rapidly and irretrievably. With rehearsal and effort, information may be consigned to longer-term storage. Whereas the short-term memories consist of mere shifts in cell function, long-term memories are thought to involve more stable changes, changes in structure. Long-term memory requires genes to "turn on," producing new proteins that are used to physically modify the shape of synapses. Like a well-worn footpath, this more permanent memory trail can now be relocated and accessed with ease in the future.

Already damage to Ma's hippocampus was beginning to interfere with her acquisition of new memories. Later, as storage areas in her cortex became affected as well, even the architecture of her long-established synaptic connections would be destroyed—with the whole garden in disrepair, even the footpaths would erode—and she would begin to have trouble retrieving old ones. This was a part of the disease I understood well, one reason I didn't want to look at the facts. But for some reason the facts had suddenly become overwhelming to me: her absolute, honest ignorance of the loan somehow broke down the last of my resistance.

I had already had so many opportunities, so many points when I could have *seen;* yet this particular gap in Ma's memory was the one to shake me loose; it lifted me just far enough out of my denial to free my mind. I overcame years of emotional entropy, my consciousness pivoted smoothly around, and I found myself facing a new direction. This was it for me. In the end, what convinced me we had a problem was not Ma's desperate need to have enough gas in the car, nor her problem with directions, nor even her inability to care for Eddie when he was dying, but the fact that she forgot a ten-thousand-dollar loan. A retired schoolteacher does not forget a ten-thousand-dollar loan.

The following Tuesday I attended a benefits fair at work. I collected balloons and buttons for Zoë, ate the free candy at various booths, had my fat content tested at the health station (a shocking 30 percent; I wasn't getting any exercise at all these days), and listened to a vice chancellor talk about the rich community I'd so recently joined. But the section of the brochure that caught my eye was listed under "Family Caregiver's Support"; they would be showing a movie about Alzheimer's disease.

I felt self-conscious entering the small room. I looked around furtively, hoping not to see anyone I knew, and sat at the very back, as though attending a self-help program for the first time. Caregivers Anonymous. Why was I there? Could this really have anything to do with me and my family? I was just about to get up and leave the room when the lights went down and the screen lit up.

The movie was Deborah Hoffmann's *Complaints of a Dutiful Daughter*, the story of a Jewish lesbian filmmaker whose mother has Alzheimer's disease. Superficially, this woman resembled me minimally; her family history was nothing like mine, her culture, her lifestyle—and yet there they were, a mother confused, obsessed, forgetting, and her loving daughter, trying to understand, trying to find compassion amid the frustration and heartbreak.

I paid close attention as Hoffmann described her mother's early symptoms, by now so familiar to me—the little memory slips, the worrying. As the film followed the inevitable progression of her mom's disease, I found myself leaning toward the screen. Deborah's mother, forced by her disease to truly live in the moment, ends her lifelong struggle against Deborah's lesbian identity, and Deborah, equally powerless to change another person's path, must abandon hope of restoring her mother to her former self. It was this admission of her powerlessness to change her mother's condition that struck me the hardest: I saw for the first time what a relief it might be to accept the facts and stop fighting.

Over the next three days I had multiple private conversations with Daniel, Alice, Pat, and the nice lady at Cal who counseled family caregivers. I described Ma's symptoms to the professional. I described the movie to my family. I listened as if for the first time to Pat's analysis and to Daniel's concerns. And on Saturday morning I finally talked to Ma.

"Ma, there's something I need to ask you."

"What is it, honey?"

We were sitting at the kitchen table. Sunshine filtered in between the redwoods and played on the pine breakfast table as we sipped our steaming coffee. Zoë was leafing through a book on the dining room floor, but I saw her attention shift to me when she heard my tone of voice. Zoë never missed a beat.

I took a breath. "Well, I need to ask you to come with me to a special clinic and have your memory tested."

"A special clinic? Whatever for?"

"It's a research center run by UC Davis, called the UCD Alzheimer's Disease Center."

She frowned.

"It's actually in Martinez, about ten minutes from your house. It's not just for people with Alzheimer's. They do research on memory and aging, and— Well, it turns out they will do a free, really thorough evaluation of anyone at all."

Ma didn't say anything.

"They are especially interested in people who are having problems with their memory."

She clasped her hands together in her lap and shook her head. "Oh, honey, I don't have a *problem* with my memory. Do you really think I have a problem?"

As always, my strongest instinct was to defer to her; she was my mom, my leader and protector. She had driven the car, corrected my grammar, taught me to cook and to ride my bike, made me buckle my seat belt. No matter how much evidence I gathered to the contrary, I still had trouble overcoming the belief that she knew better. I wanted to say, *You're right, what was I thinking? Let's just forget it.* But miraculously, the group consensus I'd reached with Daniel, Pat, and Alice protected my resolve. I stuck to my guns.

"Well, Ma, right now . . . you have been having trouble with directions around town. And you forgot Eleanor's name. And you forgot you loaned us *ten thousand* dollars." Zoë glanced up at me, cocking her head, maybe forming a question.

"Oh, posh. Did I really do that? I suppose if you all say so, I must have, but I'm still not sure I really did that."

She still didn't remember! I felt my resolve solidify. At the same time I felt bad for her. She seemed so irrationally stubborn, so childlike.

"Ma, you did that. You loaned us all that money and you forgot. Think about that. If a friend of yours did that, wouldn't you'd be worried about them?"

"Well, if you put it that way, I suppose. But I just think you're making a lot of trouble here out of nothing at all."

I felt my shoulders tense. That wasn't fair. "It's *not* nothing, Ma. I don't know what it is, but it's *some*thing. I don't think this kind of thing normally happens to people your age. Alice and Patrick and Daniel all agree. And this kind of forgetting makes you vulnerable." I was trying every kind of pressure I knew: loving-kindness, group pressure, fear tactics.

"Someone could take advantage of you," I said, my voice softening. "We just want you to be safe." As I said this, I saw her face fall, and she swallowed, looking away for a moment. I felt how true my words were, how desperately I wanted to protect her, and my eyes stung as the tears threatened to come.

"Ma, it'll be easy—the research center is practically next door to you—it's incredibly convenient. It's all free. Ma? We're not accusing you of doing anything wrong; we're worried about you. It's because we love you."

"Oh, darling, now, you don't have to worry about me. I might have some little difficulties—I'm sure I do. But Alzheimer's? Isn't that a more serious thing?"

She had mustered up some energy to say this, but I could tell her determination was faltering. Again, I had the urge to buckle, to acquiesce, but I knew I had to hold fast, to be gentle but firm. Loving but rational.

"Ma, we don't know if you have Alzheimer's. We just want to know what *is* going on, and to help you if we can. Maybe it will turn out to be nothing. But maybe there's something these people can do to help. Anyway, it can't hurt to find out, right?"

It was hard work, playing the authority figure with Ma, but as I began to patiently explain the situation to her, I realized this was already a very familiar role, gentle and persistent: it was the way I spoke to Zoë.

"I don't know . . ." she said, looking defeated and small.

"There's nothing to be afraid of. It's just a checkup, and I'll be right there with you."

"I'm not *afraid*."

I flattened my lips together, holding my mouth closed to hide my disbelief, just as I did when Zoë said the same thing. "Well, then?"

She stared at me for several seconds.

"Oh, all right."

"All right?"

"I'll go—if you and Daniel say so. But I wish you wouldn't. I think it's a lot of pish-pash."

I smiled. "So we'll note that you are going under protest?"

She shook her head at me as I moved in to hug her. "Oh, you," she said. "Sometimes I just don't know about you." Zoë got up and moved over to us, drawn in. She wanted the hug, too.

"That's okay if you don't know about me," I said to Ma as I reached her. "I love you anyway."

"Yes, I suppose you do."

The Alzheimer's clinic gave Ma an appointment six weeks away. There would be two days of interviews and testing, after which a team of doctors would compose a diagnosis together. They asked us to have her general practitioner at Kaiser Hospital order the best brain scan available under her health coverage, and a battery of blood tests. Otherwise for the next six weeks I was off the hook; I'd made the arrangements and didn't have to think about it until the time came. In the interim Zoë and I were going on a happy mission: Alice was about to give birth to her second child.

Alice's due date had come and gone, and we were fast approaching the Wednesday morning she would be induced if labor didn't begin on its own. Zoë and I were going to head up to Seattle and take little Sam while Alice and Josh were in the hospital; Pat would wait and fly up to join us that Friday night. Ma's weekend might be a little lonely, but she had Daniel in town if she needed anything.

I was looking forward to the time away from both Ma and Patrick. A few hours out in the Seattle parks and wading pools with a couple of happy three-year-olds sounded like the perfect antidote to the overwhelming grown-up issues in my life. Not only had I finally taken a concrete step toward understanding Ma's situation, but I'd been in a months-long struggle with Pat over family planning. The more I wanted a second child, the less he did. I kept bringing it up, and he kept saying it was just too early for him. It was turning into a moot point, however—he was so grumpy and I was so sad, we rarely had the desire to have sex anyway. I missed him, I resented him, I couldn't figure out how to make him understand. He was an only child, and his mom was still young and hearty; he just didn't feel how important a sibling could be. I thought maybe time away would help me gain perspective—and maybe seeing Alice and Josh with their new baby would help convince him. I hoped.

By chance the night Pat arrived to join us was our anniversary: four years. I was relieved to see him. Three full days in the Seattle parks with two toddlers had been more exhausting than I'd expected. Across the table at dinner Pat kept making jokes that Alice and Josh missed, and I kept smiling over at him, grateful that we weren't the ones to be sleep deprived and overwhelmed. I had missed his twinkle, his sarcasm, his ideas. At one point a very fussy toddler Sam was walking around with his hands in his pants. "I think he's feeling a little testie!" Pat remarked. I laughed out loud as Alice and Josh shook their heads, too tired to join in our silliness. I loved my irreverent Pat. It was our anniversary, absence had made my heart grow fonder, I was no longer the full-time babysitter and maid, and life was good.

That night I dreamed of Pat. My body met his in a warm, smooth, perfect embrace, the mere first meeting of our flesh propelling me almost to climax. Our hearts hummed, our embrace grew stronger, my breath quickened—and just then, my eyes flew open.

Pat and I were on the creaky upstairs futon in Alice and Josh's guest room, with Zoë between us. It was hot, and we had all dozed off after bath time with the covers off and the windows open. Wide awake, I turned to Zoë. She was out cold, with both arms flung up above her head and a little trail of drool running down her cheek onto the sheet. I smiled over her at Pat, whose chest was softly rising and falling more slowly than Zoë's. I loved his sleeping face—he had incredibly long lashes, and with his eyes closed, they looked even longer, almost doll-like.

I carefully pulled the unused comforter off the bed and onto the floor, then gently slid Zoë over to the edge of the bed, supporting her weight as she sank down into the soft nest with a sigh. Then I turned to Patrick and aligned my body with his. He lifted just far enough out of his slumber to turn to me, affectionate, hungry. His touch was pure sweetness, full of love and welcome, and I wrapped myself around him as tightly as I had in the dream. For once I wasn't thinking about babies; I was just thinking about Patrick. Apparently he was only thinking about me, too; he didn't speak at all, except to murmur, "I love you, Sybil," as he moved directly into me. The thing just happened, a beautiful anniversary present for the both of us.

Three weeks later, I would buy two pregnancy test kits at Longs Drugs and take them into the bathroom while Zoë was napping. It was early to be checking, but I had such a feeling about it. I would stand by the sink, light-headed and slightly nauseated, and slowly shake my head in wonder: the neat little window of each pregnancy test clearly displayed a faint pink line.

CHAPTER 11

Diagnosis

Zoë and I arrived at her preschool just as it opened, and I left her with her teacher, David, in the garden munching whole-grain organic toast before I headed east across the newly green hills of Contra Costa County. I was playing hooky from work to take Ma to her third appointment at the Alzheimer's Disease Center. The new semester had begun just as I ended the first trimester of my pregnancy, and just after Ma had completed two full days of testing. The doctors had composed their diagnosis. This was September 6, 2000: our day of reckoning.

As I drove, my mind wandered to the coursework I had been troubleshooting in the lab, an exercise in which the students would be recording nerve impulses from the legs of crickets. It was nice to be back at work after my summer off; I missed Zoë, but she seemed genuinely happy to be back at preschool, and with the morning sickness gone, my energy was up. I was getting into the pleasant groove of puttering around the lab, gathering equipment, and systematically testing parameters while listening to the local public radio station. My mind quieted during those hours in a way it never could with a toddler in tow.

When I arrived at Ma's house in Martinez an hour later, she rushed out of the house before I'd even parked.

"We'd better hurry!" she cried breathlessly.

"Ma? The appointment is in forty-five minutes, and the clinic is only five minutes away. If we leave now, we'll be too early."

"That's what you say," she said as she climbed in anyway. "I just hope you know what you're talking about."

I reminded myself to breathe as I reached over to help her with her seat belt. I had been congratulating myself for being on time, and now she was going to criticize me. I hated being in charge of my mom, coercing her to do the right thing even as she complained about it. I wanted to rebel, to return to my role as the carefree kid—the one who cut off her ponytail in a moment of liberation while hitchhiking across Europe and sent it home in a box with a few snapshots and mementos and a note asking could they please loan me more money. I wanted to throw myself down on the bed and cry, like I did when I was thirteen, and have Ma reassure me that I would find love one day. I wanted to be seven again, so she could toss me wriggling and sprawling into the pool, where I would giggle and snort until I had a nose full of chlorine. My role with Ma was to receive, not to administer; to ask her, not tell her, what to do. But that wasn't the case today. Ma needed me today. She needed my expertise, my encouragement, a push here and there.

"Well, we got a beautiful day for it, didn't we?"

"Oh, I suppose so," she said, shaking her head, "but I still think this is all a big waste of time. What do they think they're going to find out from all those silly tests, anyway? Probably nothing."

"Well, maybe so, Ma. That would be great, wouldn't it?"

She didn't answer but looked at her watch and sighed loudly.

"Now, Ma, we have plenty of time." I'd intended to use a reassuring tone of voice, but it came out sounding more like a scold.

Suddenly she lunged forward and began to paw frantically through her purse. "Oh dear," she said, "oh no, oh dear . . ."

I silently counted to ten.

"Oh, thank goodness!" she finally said, sighing, holding her driver's license to her chest. "I can't go out without this."

When had she begun to misplace every little thing? I wondered. When had this worrying begun? I recalled accusing her of being a "worrywart" when I was ten, but that had been a different kind of worrying. Those had been a loving mother's superstitious "don't forgets." Don't forget to wash your hands; don't forget to look both ways; don't forget to fasten your seat belt. Now she fretted about what to order for breakfast, about the milk Zoë spilled on the kitchen table, about whether her TV program would really come on at nine. These days her worries seemed to stem from general angst rather than a desire to ward off real danger. And now her "don't forgets" were more often addressed to herself, in the form of little reminders she had begun to post all around her house: Take vitamins. Take shower. Do laundry. Bring ID. Don't forget.

As we approached the entrance to the hospital complex, she kept looking anxiously out the car windows to either side.

"I sure am glad you're driving," she said. "This is a hard place to find."

"It's the same place we went both times for the testing."

Sometimes I turned mean when confronted with Ma's waning navigational skills. When I was at the wheel, I knew if I took a street that was not on her chosen route, she would begin to twist and turn in her seat, checking street names and searching for landmarks. "Are you sure this is right?" she might ask. "Yeah,"

I'd say breezily. "Oxford is better this time of day; there's too much traffic on Shattuck." I didn't reassure her or offer any tips about where we were. Cruelly, I would watch her in silence as she agonized over the unfamiliar territory, and then study her reaction as I returned to her usual path, waiting to see how long it took her to regain her bearings and feel comfortable again.

There was an element of testing her in this: How badly limited was she? How far off the beaten trail did I have to go before she felt lost? There was also an element of distinguishing myself from her, proving that I was flexible, capable, and not subject to whatever affliction she was suffering from. When I did this, I was catty, sharp, and cold; I didn't have to feel powerless, I didn't have to feel sad.

"This doesn't look familiar," she said. "Are you sure this is right?"

"Yep." As I slowed at the sign for the Alzheimer's clinic and signaled to turn, I silently observed that she was disoriented in a setting that should be familiar to her. Hair: greasy. Same clothes as yesterday. I had developed a sort of internal accounting system to catalog her subtle personality shifts and memory lapses. *Pathological or normal?* I asked myself. Pathological, I decided in this case—but then at the T, I hesitated. Left or right? An instant before, I'd been sure of the directions, but now I wasn't sure of the way, either. Was that just a stress-induced brain freeze, or was I having memory problems myself? I turned right, and around the next bend we saw the familiar low building surrounded by institutional shrubbery. It was my turn to sigh my relief.

While we waited in the reception area, I tried to engage Ma in conversation: "How was your walk this morning?"

"Oh, fine."

I waited for more, but that was all she had. I tried again: "Anything new in Daniel's life?"

"I suppose he's fine."

"Are you nervous?" I asked.

"Oh, I don't know."

Exasperated by her refusal to engage, I dogged her: "It's okay if you feel anxious, you know."

She scowled. "I'm fine. I don't know why you had to bring me to this place."

I descended into a familiar vortex of guilt and self-doubt. I doubted my own authority. In spite of the fact that I had degrees in both psychology and molecular biology, and a much better than average understanding of how degenerative neurological diseases affect behavior, long habit dictated that I defer to the expertise of my parent: Mother knew best. I doubted my judgment. Had I mistaken perfectly normal behavior for an aberrant lapse? I also questioned my motive. Maybe I was just looking for any excuse to get my old mom out of my hair so I could reclaim my life. Ungrateful, punishing daughter. God, what if I was wrong and this was just normal aging? I would have dragged poor Ma through all this grueling testing for absolutely nothing. But by now I knew to return to the facts: She barely smiled anymore and almost never initiated conversation. She had forgotten her best friend's name. She had forgotten that she'd loaned us ten thousand dollars. I was doing the right thing; I had every reason to worry.

I had been reading up on dementia over the past several weeks, peeking at current articles online. The word *dementia* comes from the Latin *de,* meaning "apart" or "away," plus *mens,* or "mind." It is a catchall term for any progressive decline in cognitive function, at a rate that surpasses the medical establishment's definition of "normal" for one's age. By far the most common cause of dementia is Alzheimer's disease, and Ma's recent behavior certainly fit the bill for the beginning of this progression: She already had minor forgetfulness, she had begun to avoid new situations, and she got lost when she strayed from her home territory. She rarely initiated conversation, had trouble paying attention, and was often extremely indecisive. In fact, all the new traits that so challenged my idea of who she was turned out to be hallmarks of this disease, this progressive and debilitating, incurable disease.

I reminded myself that there were other, less common conditions that could bring about some similar symptoms. One was controversial, because it was both poorly studied and a fun-spoiler: alcohol-induced dementia. I had no doubt that Ma's drinking had affected her thinking, but I wasn't sure if the symptoms I had read about fit with Ma's: in addition to memory problems, severe alcoholics often had poor coordination, trouble executing voluntary movements, and difficulty recognizing common objects. Ma wasn't suffering in these ways, that I was aware of. Another form of dementia, called vascular dementia, was caused by hypertension: the patient suffered a string of tiny, imperceptible strokes that caused cumulative brain damage, often resulting in memory problems and loss of executive function (the ability to plan, strategize, initiate; the ability to use one's head). Underproduction of thyroid hormone and certain vitamin deficiencies could also cause memory problems and neurodegeneration.

All these less common dementias had the advantage that the root cause could be addressed medically. The symptoms of Alzheimer's disease could be treated for a while, but its cause could not. I hoped against hope that the doctors would tell us Ma had one of the less horrible conditions—but based on everything I had read, this sure looked like Alzheimer's.

For so long in school I had focused on how neurons worked; how over a lifetime they found their proper places in the brain, formed networks, and built memories. Now I was considering the other side of the equation, the gradual decline of this system that had functioned so smoothly for over sixty years. I wondered what fragile neural structures might be shrinking with each subtle change in my mother's character.

I imagined the inside of Ma's brain decades ago, when she had been well: ripe, thick, strong neurons, supple and elastic, their tendrils forming a complex and highly ordered web within her brain; axons thriving like young vines in a jungle, glistening, growing, and twitching with activity: sparks running the length of them, flashing bright colors onto the dark, moist forest of cells around them. The strong, flexible limbs of her neurons ran in every direction, all teeming with information, while clusters of cell bodies, knotted into nuclei like teams of top executives, strategized and analyzed the signals, integrating them into coherent thought.

The neuronal outgrowth and pruning that had occurred over the decades of her life had enabled Ma to walk through her day being Ruthie; they had defined her character and coordinated her thoughts as she camped with her cowboy father as a child, headed off to college and married as a young adult, and then gave birth to my sister and me. As she signed petitions, marched at peace rallies, cut out fall leaves from construction paper, and lovingly fixed them to the bulletin board of her first-grade classroom, every event was physiologically recorded, the configuration and activity of cells in her brain finding and maintaining that particular inner shape that made her who she was.

There would have been a momentary electrical lull during my parents' Friday night cocktail hour. The alcohol in her blood, which had easily passed into her brain, inhibited the release of many neurotransmitter chemicals, slowing signals, preventing communication between the brain's constituents. I pictured the colors shifting from energetic bright yellows, oranges, and reds down to cool purples, greens, and blues, how the mist thickened, the sparks became fewer, the ropes of neuronal processes hung slack. But the next morning, with most of the alcohol metabolized, the coffee drunk, and a new set of receptors stimulated, a new cascade of activity would kick in, and she'd be back to normal.

Until the Alzheimer's, if that was what this was, came into play. One day certain neurons, maybe only a handful out of her one trillion, would shoot bursts of current down their axons to no avail; no response would be possible. Not enough neurotransmitter was released, and so when it crossed the synapse, it could not excite the next cell down the line.

At first, only a scattering of connections would be lost, and the consequences would be so slight as to be undetectable. At the earliest point in the disease, recalling events from the past, Ma would be unaware of any change. Remembering a trip to the coast six months earlier, she would know perfectly well what had happened and how, yet she might feel something nag at her subconscious mind. All the elements would be in place—the car, the hotel, the gift she had brought back for her next-door neighbor— yet some subtle quality of the memory would have shifted ever so slightly, the colors and textures of the ocean that had touched her so deeply now somehow dulled.

Early on, statistically, so little would change that she would not be consciously aware of any shift, nor would anyone else, but inside there would be trouble. Neurons would begin to secrete large amounts of a protein called amyloid, which would proceed to gather in the form of hard, insoluble plaques between the cells of her brain. Meanwhile, her cellular transport systems may have already begun to fail. The inner structure of the neuron is strung with protein tracks like highways through the jungle, along which minuscule molecular motors truck important molecules, including neurotransmitters; they continually traverse the cell like so many tiny delivery trucks. But the delicate internal scaffolding of the cells' graceful slender branchings would now become twisted and disorganized, clotting together to form tangles made of a protein with a beautiful name: *tau*. Cargo sacs carrying cellular goods would begin to pile up in massive traffic jams. Some cells would die, leaving behind tangles of tau as markers of their demise in the tissue of Ma's brain.

Over several years' time the amyloid plaques and tangles of tau would continue to accumulate, neurons would start to die in greater numbers, and Ma would begin to change. Now not only would the quality of her memories change, but she would forget some recent events altogether; she would miss appointments, forget to bathe, and lose her way.

Eventually, Ma's brain would become a bleak wasteland where the lush forest had once grown, littered with sticky plaques and tangles of tau like piles of bones. Eventually, Ma would not be able to speak, recognize her loved ones, or control her bowels and bladder.

I looked over at her, now hunched in her chair, staring down at the beige carpet, and for an instant I thought I knew: it was Alzheimer's. Then I told myself I couldn't possibly know, I had to leave it to the experts. I surreptitiously reached down and patted the sides of my belly. This time I had swelled up much faster; I looked more like six months pregnant than three. For a moment I silently telegraphed my love to the tiny sprout inside, as if it could catch Alzheimer's, as if this were a threat to its well-being. Then the neuropsychologist appeared, looking crisp, slim, and far too young to be a doctor.

"Ruth?"

Ma looked up but didn't answer.

"Yes," I replied, smiling, "we're Ruth."

The doctor returned my smile as she led us into a small

conference room with a long white table. I spread my notebook and papers out before me, the picture of organized, knowledgeable competency. At the opposite side of the table the neuropsychologist sat down next to the clinical nurse, Carol, who reached over to shake our hands, greeting us warmly by name. Carol was rounder and rosier than the neuropsychologist, and I felt immediately comforted by her presence in the room. Ma looked up, expectant. She briefly met my eye. She seemed suddenly less resentful of me, with the medical staff on one side of the table and us together on the other. We made small talk as we waited for the third member of their team, a neurologist. I felt a brush of tiny butterfly wings inside my belly: not nervousness. I folded my hands across my abdomen with a tiny private smile.

I was wired and anxious but not afraid; I just couldn't wait to hear the bottom line. I knew something real was happening to Ma, something I didn't like and didn't fully understand. I needed to move out of this squirming, nebulous place where I saw the changes in her but could not name them. If this turned out to be Alzheimer's, at least that was concrete information, something to work with.

That was not all, though. I had to admit that a certain hidden part of me also hungered for calamity; it wanted to be shocked and appalled. A diagnosis of Alzheimer's would be sensational, and like a driver passing an accident on the freeway, straining in spite of myself to see the spray of blood on the windshield, I was actually eager for the thrill of the unthinkable.

Between my need to pinpoint a cause for her behavior and my sick desire to hear the worst possible diagnosis, I was nervous, but I didn't feel as much real and present concern for my mother as a daughter ought to feel at such a moment. Still, I reached for Ma's hand. I did this not so much out of an immediate need to warm or comfort her, the way I would reach for a crying child, but because I had seen compassionate people portrayed on television at critical moments like this one, and this was how they behaved. As Nurse Carol smiled reassuringly across the table at us, I guessed that we looked correct; Ma looked scared, and I looked supportive. In truth, she was annoyed and I was excited.

Under Carol's caring gaze, I felt a swell of remorse. My mind flooded with memories: Ma's young smile, her summer freckles, her soothing voice; Mommy kissing Daddy, reading to Alice and me; the feel of her strong arms as she picked me up and hugged me fiercely. I didn't want this to be the end. I just wanted my old mom back.

The door opened, and the neurologist entered: he was tall, around my age, with dark, empathetic eyes and a weary but friendly smile. He was wearing the white lab coat that signaled his Doctor authority. *Good,* I thought. Despite all her feminist ideals, my mother still really listened best when a man spoke. He sat. But after shaking our hands, he turned and nodded to the neuropsychologist, who began to speak, her eyes on her papers. There was no meaningless preamble or strained apology, like in the television hospital shows. Nurse Carol smiled gently across the table at Ma as the neuropsychologist said simply, "Ruth, we believe that you are suffering from Alzheimer's disease."

I looked at Ma, who was looking down at her lap. I kept my body perfectly still, as if that could stop time. A car started outside. Finally, I nodded, took in a breath, and slowly let it out. Ma had not looked up. Her expression did not change.

"We're basing this diagnosis on several factors," the neuropsychologist continued. "Your symptoms started out subtly and then progressed more quickly over a period of time. These symptoms include problems with your memory as well as with various other intellectual processes typical of Alzheimer's patients. In the evaluation we did here last month, we put you through a battery of memory tests that helped us to quantify the extent of the memory problems you have been experiencing. Then we looked for any possible reversible or curable causes for these problems. For example, in the tests we ran on your blood, we looked for vitamin deficiencies, thyroid abnormalities, electrolyte imbalance, and infections. These tests all came out normal."

I hurried to finish the note I was scribbling before she started talking again. I wanted to get all this down, so I could report every detail to Alice that night on the phone. The neuropsychologist regarded me sympathetically. "Next week we'll be sending out a letter to both of you and to your mom's doctor at Kaiser, laying out everything I'm saying here."

"All right." I swallowed, reluctantly putting down my pen. My tongue felt thick, and a wave of nausea had just rolled through me; I didn't know whether it was morning sickness or shock, but I pushed it back. I reminded myself to be thorough, to be Ma's health advocate; Alice would have done the same.

"So," I said, "you ruled out some of the reversible causes for this kind of memory loss?" I turned to the neurologist. "What about strokes? Isn't there a kind of dementia caused by a bunch of tiny strokes?"

The neurologist nodded. "That's right," he said, "but we use imaging techniques to detect physical damage from the strokes. Your mother's blood pressure is good, she has no history of stroke, and in any case the CT scan they did at Kaiser revealed no damage. So we ruled out that form of dementia."

"Okay," I said.

The neurologist glanced over at the neuropsychologist, who continued. "In situations like this, when a person has progressive loss of cognitive function, observed by the patient and family members as well as through the kind of neuropsychological tests we put you through; when we see a slow onset and gradual progression; and when lab tests come up normal for other causes, we make the diagnosis of Alzheimer's disease."

Finally, Ma reacted. I heard a small rush of air as she let out her breath. Her hand was hot and damp under mine, and her whole body suddenly sagged for a moment, like a deflated balloon. Then she looked up at the strangers across the table, incredulous. "So I really have it. I have Alzheimer's?"

I squeezed her hand, this time out of sincere and immediate empathy. For the first time I heard fear in her voice, a high breathy waver. And shock. I realized then that she really hadn't believed it. She had just not realized the severity of the difficulties she'd been having.

I was one step ahead, having suspected it all along, but it still felt unreal to me. I needed to be absolutely sure. The scientist in me stepped in. "You said you'd ruled out other causes," I said, "but statistically, what is your confidence level?" I realized I was speaking in a language Ma might not relate to. "How sure can you be?"

"I understand your need for concrete answers here," said the neurologist, "but unfortunately, we can't be one hundred percent certain of our diagnosis without an autopsy. On the other hand, most research studies show a high rate of accuracy of diagnosis compared to autopsy in the hands of experienced clinicians."

I didn't hear the rest of his explanation. What I had heard was that they could give us a concrete answer, but she had to die first. The juxtaposition of the words *diagnosis* and *autopsy* had sent me reeling back to the darkened hospital room where my father had died of lung cancer, back to Eddie's deathbed. Ma had lived through their deaths; she knew what a slow and painful end looked like. I searched her face for signs that she was remembering, too. Her expression hadn't changed, but she sat forward and asked simply, "But how bad is it?"

We all paused blankly. How bad is Alzheimer's? I wanted to say, *It's bad! Really, really bad. First you slowly lose your mind, then you are moved to a home where you sit drooling and peeing yourself for a few years, and finally, once you and possibly your family are absolutely broke, you die without ever getting a chance to say good-bye.*

The neuropsychologist understood what Ma needed. "The estimate of how far along you are is going to vary depending on who you are talking to," she said. "Here we rank the progression of the disease as early stage, middle stage, or late stage, and we speak of the severity of the symptoms as mild, moderate, or severe. You seem to be in the early stage with mild impairments."

I nodded my approval, smiling with my eyebrows raised encouragingly, as though Ma had just gotten an A on her report card.

"But what do I *do* about it?" asked Ma, flustered. "Is there a medicine I take to make it go away?"

But she must know there's no cure, I thought; *Alzheimer's is in the papers every day. Was she just in extreme denial, or had the Alzheimer's already erased the facts of Alzheimer's from her memory?*

Nurse Carol looked at my mom. She took time to make sure that Ma was engaged. "There's no cure for Alzheimer's disease," she said earnestly, "but we have every indication that you're only in the first stages of the disease. You're only sixty-nine, and in good health otherwise, and in most cases Alzheimer's progresses over many years. I've worked with dementia patients in the field for a long time, and the quality of life now for people with Alzheimer's disease can be very good."

They did recommend one medication to Ma, a pill called

Aricept. As Carol had said, it was not a cure. It was a drug designed to combat memory loss in the first stages of the disease, by compensating for the shortage of a neurotransmitter called acetylcholine (ACh for short).

Neurotransmitters like ACh are the messengers that cross the synaptic gap between the axon of one neuron and the dendrite of the next in order to pass information from cell to cell. A signal in the form of electrical current travels down the axon of the first cell until it reaches the tip of the axon, which is separated from the surface of the next cell by a very small gap. The arrival of this signal at the axon's tip releases a batch of neurotransmitter into the gap. This chemical diffuses across the gap and washes up on the shore of the receiving cell's dendrite, where its arrival causes a new current to start up in the new cell. In turn, this new impulse shoots down the axon of the receiving cell to any neighbor it is synaptically connected to, and so on.

This neurotransmitter-mediated process is essential to every sensation, every command the brain sends to the muscles to move, every involuntary blush and blink, and every single memory. A nervous system with a shortage of neurotransmitter is like a corroded circuit board; its connections become intermittent and unreliable. If enough cells lack enough neurotransmitter, nothing remains the same.

As soon as its job is done, ACh is normally removed—either taken back up into the neurons for reuse, or broken down by enzymes in the synaptic cleft—and Aricept's job is to stop the enzymes that normally break down ACh. (These enzymes are called cholinesterases, so that makes Aricept a *cholinesterase inhibitor*.) Thus, in the presence of Aricept, ACh is not removed from the synapse as quickly; what little ACh the deficient cells still produce is preserved, so that it might accumulate in sufficient quantity to produce a signal.

Unfortunately, Aricept normally works for only six to nine months, in the early to middle stages of the disease. As time passed, not only would Ma's neurotransmitter stores become increasingly depleted, but more and more nerve cells would die outright; eventually, there would be so few synapses, and so little ACh in them, that the medicine would become completely ineffective. A medicine that prevented the cells from drying up and dying in the first place would have been preferable, but such a medicine did not yet exist.

At first I let myself be soothed by Nurse Carol's practical tone as she patiently explained to Ma that Aricept might help her to remember better in the first year, and if she was feeling depressed, they might also prescribe some antidepressants, but that was really up to us. She helped me feel like I didn't have to worry for the moment about Ma becoming a drooling, incontinent lamb who would wander out into the street every time I left her unattended.

"However," I heard her say, "it is important to plan now for the time when you will need more assistance. Alzheimer's disease does worsen over time. The progression varies a lot from person to person, and it's impossible to know whether in your particular case things will go quickly or slowly. The bottom line is that you will eventually need the kind of care that family members cannot give you at home." She turned to me. "Probably the most important thing you can do now is to plan out how you are both going to get your needs met in the future."

Oh. Scratch that business of not having to worry yet. I glanced at my mother, who looked poised for action, with weird glassy eyes, like a dog at the pound when you touch the gate of its cage. I hadn't seen her pay this close attention to anything in months.

I pressed the doctors hard on the issue of time, as I imagine many people do. I wanted to know the bottom line. How long until she forgets my name? How long before she begins to have physical symptoms? How long until she dies?

I couldn't ask these questions outright, with Ma sitting right there, though. Out of habit I used a technique Patrick and I had begun to employ to communicate adult subject matter at the dinner table in front of Zoë. "It would be pleasant to interact with you in the manner of mature and consenting adults later on this evening," one of us might say with a grin. When we did this, Zoë would stop eating and look up. She would scrutinize my face, then his, her mind working double-time as she tried to parse the sentence she knew was packed with meaning. Here I inquired opaquely, "Can you make a prognosis as to the time course of the progression of the deficiencies brought on by the dementia?" Ma's face remained blank.

"We hesitate to guess," Nurse Carol explained, "since individuals can progress at very different rates. It can be five years or it can be twenty, but the average span from the time of diagnosis is about ten years."

Ma looked at me. "Ten years until . . . ?"

Nurse Carol said firmly, "Ten years on average for the disease to play itself out, but for many people it is much longer." I could tell that Ma was not satisfied with the phrase "play itself out," but she didn't ask.

I didn't, either. Play itself out, indeed. I could imagine how it would play itself out for me. The baby had begun to move again, signaling to my full bladder that this meeting must end soon, and I thought, *This is my future. My two children will grow as my mom disintegrates, and I will be there in the middle. I'll nurse the baby; I'll find a nurse for my mom.* I imagined my life as a great flurry of doctors' appointments, teacher conferences, estate planning, form signing, and bed making. A few minutes later, as we left the center carrying a sheaf of information about caregiving and

Alzheimer's, I developed a screeching headache, as though my cranium had needed to physically stretch to accommodate all the new information.

That night after Zoë fell asleep, I told Patrick I had to go out. I was so overwhelmed by the consequences and implications, I needed space to expand, to explore my changing role in the constellation of my family. I drove into the regional park in the hills above our house to Lake Anza, a frequent summer bathing spot, and seeing no park rangers, I walked by moonlight along the footpath through the warm night until I was halfway around the lake. On a rock outcropping under the stars, I slipped off my clothes and then I stepped into the black water. With a quiet slurp, it swallowed me up.

Surfacing, I turned on my back and watched wisps of high fog run past the half-moon as though they were in a hurry to get somewhere. The agony of not knowing was over; I could move on to whatever came next. Floating there with the faint moonlight on my pregnant belly, I felt a new forlorn, soft something unfolding, and after a moment's confusion, I recognized it as compassion. It wasn't Ma's fault. There was a good reason why she was acting like a stranger; she was changing, and she had no control over the changes. She had a disease. I had been so afraid of the growing separation between us, but now for the first time I was able to see it in a positive light. Now, in distancing myself enough to know the problem and confront it, I might become of service to her. I made a vow to be kind to the worried stranger who inhabited my mother's body. I made a wish for the little life within mine. Then I swam back to shore.

CHAPTER 12

Impulse Control

I squeezed my large belly past a man with an even larger one at the doorway of Saul's deli, and Zoë, Ma, and Pat all followed me outside. The sunny sidewalk was full of kids, dogs, and young couples with mussed hair. We had arrived ahead of the Saturday morning crowd for our bagels and eggs, but by now there was a line out the door. Patrick immediately headed for the street, newspaper in hand. He had been sweet with Ma over breakfast— they still argued over the bill every time, although the game had become so formulaic, it had lost some of its joy—but these days he seemed perpetually frustrated by the chaotic combined energy of inquisitive Zoë, bumbling Ma, and big, round Sybil the troop leader. I watched him withdraw, wishing he'd turn and help me as I brought them to the car. Zoë tugged on my pants leg, trying to use me as a blocker for the large German shepherd tied to our parking meter. Ma started to follow Pat but then turned and hovered between us uncertainly.

"We've gotta stop in at Long's on the way home," I announced. Then on second thought I looked down at Zoë. "Do you have to go potty before we leave, sweetie?" Zoë shook her head. I looked at Ma. "Do you?" She nodded sheepishly.

In the five months since Ma's diagnosis, I had begun to expect moments like this. As long as I was careful not to become patronizing about it, much of what I had learned about parenting a three-year-old transferred well when dealing with the new Ma. Early Alzheimer's was a tricky disease, coming and going. There were whole weeks when she was bright and joking and normal-seeming, but then she would suddenly become vague and bewildered again, have trouble holding up her end of a conversation, and forget what she'd just been told.

I knew this waxing and waning was normal and expected. It might reflect the redundant nature of the nervous system—we start out with more neurons than we need, so when one dies, another can sometimes "learn" to function in its stead—until that cell, too, is affected. Alzheimer's disease also reduces the levels of key neurotransmitter chemicals in the brain, and these chemicals are known to fluctuate with diet, mood, and activity. Neurons with dwindling neurotransmitters, already hovering at the brink of dysfunction, may be more sensitive to small variations due to a dark day or a good meal, causing the behavior of the afflicted person to fluctuate as well.

I understood this, and still it was impossible to get used to. I got so bewildered when the old Ma suddenly popped up; I didn't know how to react or how long she'd be around. Sometimes I was so happy to see her that when she lapsed again into forgetting, fretting, or the blank stare at the dinner table, it felt like a double betrayal, a double disappointment. She'd begin to act just like a tired kid, asking me the same question three times in a row, or forgetting to go to the bathroom, and I'd want to scream at her to act her age.

At this moment, however, as Ma and I walked back into the

crowded restaurant, leaving Zoë and Pat to wait outside, something new was about to happen. We were making our way along
the aisle to the bathroom, past the waiters' station with its coffee-
pots and napkins on one side and a row of two-person booths on
the other, when two of my students waved to me from a nearby
table. "Hey," I said, giving them a brief, preemptive "no time
to stop" wave and flash of a smile. I was aware that they were
checking out my mom and my pregnant belly very carefully. My
students had begun to ask a lot of personal questions as I'd grown
big. By and large it didn't bother me; I enjoyed the attention. But
I felt especially pressured to do a flawless job at work; some of
the guys had begun to joke about my "delicate condition," and
people kept telling me to sit down when I was in the middle of
something. I didn't want anyone to think I couldn't handle the
job. With my students' eyes on me, I was already a little self-
conscious when Ma's purse strap broke.

At first I didn't realize why she was pausing there in the
middle of the aisle, grasping at her arm, but then the purse fell,
turning on the way down, and all her belongings spilled out onto
the floor with a crash. "Oh NO!" Ma cried, dropping down to
her knees.

"Oh God," I said, turning to Ma. "Don't you hate it when that
happens . . ." Then I saw her face: her lips were a white line, and
she was clenching her teeth, her eyes narrowed to slits. "SHIT!
SHIT! God DAMN it to HELL!" she yelled. "That was so STU-
PID! FUCKing purse, I KNEW I should have gotten a new one,
that was stupid, stupid STUPID!"

I stopped in my tracks, blushing slightly, and glanced around
helplessly as several heads turned. My students' faces registered
both concern and intrigue, but I turned my back to them, fo-
cusing all my attention on Ma. A young waitress and I both
quickly squatted down next to her, and as we scooped up the

loose change, a ChapStick, a packet of tissues, we exchanged a glance—the waitress smiled nervously, and I raised my eyebrows in an embarrassed shrug. Just as quickly as it had arrived, Ma's tirade began to die away. "I don't know what I was thinking, bringing this stupid old thing," she was saying, shaking her head. She sat back with the purse in her lap, and suddenly she seemed very tired. As she looked up at me, and then at the people around her, her face went pink, and she wiped some sweat from her forehead. "Oh dear, I'm sorry, darling."

"Come on, Ma," I said. "I think we have everything." I turned gratefully to the waitress. "Thank you so much." Ma started shakily off in the direction of the door, and I steered her gently back toward the bathroom. "It's okay, let's just get you done, and we can go home." I continued to avoid turning toward my students; if they were staring, they really shouldn't be, this was none of their business.

I was so glad Zoë hadn't needed to go. Zoë knew Gram "had problems remembering" because she was "a little bit sick." She had heard me talk in a general way about Alzheimer's disease, but not the private conversations, not the anguished best-friend heart-to-hearts. Those happened late at night or outside her earshot. When Zoë asked a question, I answered honestly, but with more optimism than I really felt. I said, "Gram forgets more things than we do, so we have to help keep her on her toes." I said, "She worries when we're late, so let's make sure we pick her up on time." I didn't tell Zoë that Gram was not the same mother anymore, that she had once been strong and bright, rebellious and independent, but had become simple and passive, fretful and agitated. When I said, "Gram forgets," I didn't elaborate that she had to write herself a note for everything she did, and then she lost the note. I didn't mention that I was not sure she was safe alone in her house anymore, or that she would gradually lose herself in a terrifying reduction of her abilities and her senses. I didn't say she was slowly dying.

"She may begin to have bursts of anger or sadness, or even inappropriate sexual behavior," Nurse Carol had warned me on the phone the week after her diagnosis. By now I had read up about this. Alzheimer's disease causes damage to the frontal lobes of the cerebral cortex, and that damage can cause a person to experience "disinhibition," the inability to suppress impulsive behavior that we normally keep in check in social situations—anger, sexual desire, sadness, fear.

Lobes are simply rough divisions or broad areas of the brain. There are four pairs of lobes: frontal, parietal, temporal, and occipital. The frontal lobes are located at the front of the brain, directly behind the forehead. They and all the brain's lobes have a very thin layer of cells at their surface called neocortex, from the Greek *neos*, or "new," and Latin *cortex*, or "bark of a tree." This is the most recently evolved portion of the brain, and it is responsible for a stunning array of higher brain functions. It consists of gray matter, mostly neuronal cell bodies and synapses, as opposed to the white matter portions of the brain, which are largely made up of axons (white because they are insulated by myelin, which helps speed messages from one brain area to another). The two roughly symmetrical frontal lobes make up the single largest cortical region in the brain. Impulse control, sexual behavior, social skills, judgment, language production, working memory, movement, problem solving, and spontaneity all rely on the frontal cortex. No wonder damage to the frontal lobes can make a thing as simple as a purse strap breaking nearly impossible to handle appropriately.

When Carol mentioned disinhibition, for some reason I had focused on sexual inhibition. I'd thought of Pat's grandma Grackle, who by the time I met her already suffered from dementia due to chronic drinking. She went on and on about her "boyfriend" Willie Nelson, and she had definitely lost some inhibitions; while amiably chatting with an orderly in her hospital, she might reach out and pat his behind, or give his testicles a friendly squeeze. But Grackle had been a bit of a wild card well before her nursing home days. I couldn't imagine my mild-mannered mother behaving in this way.

Maybe in her life Ma had needed to inhibit a different set of behaviors, so she disinhibited differently, too. Take the swearing. Both my parents swore freely when I was growing up, and so did Alice and I. It was the 1970s, we were a pack of liberals, and my parents never censored our language in the house—though they did have the sense to teach us to behave in public. But Ma didn't swear in her first-grade classroom; none of us were allowed to use our usual foul language in the homes of more conservative friends; and in the presence of Ma's parents, we had to inhibit that habit severely.

Grandma and Gramps lived in Laguna Beach, and every summer Ma, Alice, and I took a road trip south to visit them. Gramps would have been fine with the swearing, I'm sure of it—he was a cowboy, after all. Gramps was the fun one. Mornings, he chased us down the hall holding the orange juice pitcher over our heads, threatening to tip it. We'd squeal and giggle helplessly, scrambling to escape to the safety of Ma's arms, while Grandma looked on. I had seen Gramps sneak up to the garage midday to drink a beer with the guys from the construction site across the street while Grandma prepared for a church meeting, and those guys certainly didn't keep it very clean—but my conservative, Christian Scientist, tsk-tsking grandmother was another matter.

Grandma put a damper on all our fun, but we had an unspoken understanding about that. We'd drop her off at her hair salon, or her Daughters of the American Revolution meeting, and then Gramps would relax and hang out with Ma and us kids. We'd head straight to the main beach to see what color the life-guard's flag was, and Gramps and Ma would sit on a blanket and talk for hours while Alice and I played in the waves.

When Ma was a child, she and Gramps had hiked and fished together in the Sierras. I treasured the old black-and-white photos of ten-year-old Ruthie, triumphantly holding up a rainbow trout on the line, with young-man Gramps's arm slung proudly around her shoulders. On our trips to the beach, I knew they were bond-ing there on the sand, sharing that love of the outdoors that he had passed to her and she to me. I knew they could really talk there in a way they couldn't with Grandma around. Evenings we'd have to return to Grandma's world, with stuffy car rides to boring restaurants with great-aunts and -uncles who wore a lot of stinky perfume and aftershave and literally pinched my cheeks. I found those excursions so deathly, I didn't even have the heart to swear.

To prepare ourselves for these visits, all the way down the coast in the car, Ma, Alice, and I practiced our etiquette. We played a game in which we lost a nickel for every swear word we used, and we earned one each time we caught someone else in the act. "It's so fuckin' hot," I'd complain, and Alice would imme-diately shout out, "GOTCHA!!" Looking up, I'd see Ma's blue eyes laughing in the rearview mirror. "Ahhh, she got you! Pay up!" Since Ma was best at the game, there was a special pleasure in catching her; it was a nickel well won. "Dammit," she'd say as she missed a turnoff, and Alice and I would bounce in our seats, shouting, "GOTCHA GOTCHA! No, I got her first! *I* did!" As an adult, when I got cut off in traffic, all the cuss words that were part of that family culture still came streaming back to me. When I had passengers in the car, I mostly managed to keep it in check, but I could just imagine how I'd sound when I got Alzheimer's: just like Ma at Saul's.

Would I be that angry, though? Ma's anger had chilled me. I hadn't known she was capable of such fury. Had it been there all along and simply been unleashed by the disease, as the name *disinhibition* implied? How strange to think that even the sweetest-tempered person among us might carry a huge reservoir of raw hostility that she must continually dam in order to maintain her serenity. I wondered if there was some natural advantage to having an emotional cauldron bubbling with bile on one's psychic back burner. Was anger, held in check and carefully titrated, useful in the wild? Maybe it would be handy to have rage on tap in case one suddenly had to spar for the position of alpha beast.

And in the context of Ma's unwild life? I didn't know what the advantages would be, but I did know that teaching had been difficult for her in the last years of her tenure at Pittsburg Elementary. Ma had taught some pretty troubled kids then. I wondered now why she had retired early. Might she have had an early bout of disinhibition and cussed out some poor misbehaving student? I hoped not, but it was impossible to know when exactly the dementia had begun, or in what subtle ways it had affected her decisions then.

The Monday after Ma's outburst at Saul's, I visited the bioscience library at Cal. It was in another wing of the Valley Life Sciences Building, where I worked. I climbed the spiral staircase past a full-size replica of a *Tyrannosaurus rex* fossil, and I thought of Zoë. Whenever we dropped by my work on the weekend, she wanted to visit the crayfish in their tanks, ride down the long dark hallways on my swivel chair, and go to the library to see the dinosaur fossils. I didn't come down here often during the week. I was feeling a little sheepish about using work hours to research unrelated material, but I needed to find out more about the frontal lobes and disinhibition.

As I waded through research on the frontal cortex, I found

many articles about another form of dementia called frontotemporal dementia, or FTD. The name made me think of the flower delivery service called FTD; I pictured the winged Mercury of their logo delivering bouquets of lilies to dementia patients. But FTD was a serious matter. It turned out to be the second most common form of dementia after Alzheimer's.

Alzheimer's disease begins near the hippocampus and eventually extends into every lobe of the cortex, whereas FTD is confined to the frontal and sometimes temporal lobes. Once known as Pick's disease, FTD is progressive and debilitating like Alzheimer's, but it is less common (FTD is responsible for about 20 percent of dementia cases, whereas Alzheimer's claims almost 70 percent), and the symptoms are slightly different. What if Ma had been misdiagnosed? I wondered. What if she actually had FTD?

Frontotemporal dementia is an umbrella diagnosis for a complicated collection of dementias, some of which have a more frontal and some a more temporal origin. Language deterioration, including trouble speaking and understanding speech, is more typical of the temporal form of FTD, which is also known as semantic dementia. That didn't sound like Ma. The more frontal kinds of FTD chiefly impact the frontmost portion of the frontal lobe, directly above and behind the eye, called the prefrontal cortex. Prefrontal cortex is one of the most complex, highly developed neocortical regions in the human brain. It serves as a massive association complex, with incoming and outgoing connections to all the other neocortical regions, and it is important for working memory, personality, spontaneity, and "executive function," meaning judgment, problem solving, and initiative. Confusion, apathy, disinhibition, and trouble initiating, planning, or handling change are more typical of frontal dementia. That sounded like Ma.

These FTD dementias are more common in women and tend

to manifest themselves earlier than Alzheimer's disease, usually in the patient's sixties. As I read, I compulsively categorized Ma's symptoms. I thought of the countless times she had sat mute while I tried to jump-start a conversation with her; I remembered the distress that had shown so plainly on her face when a trip to the grocery store had been derailed because Pat and I decided to stop to check out a yard sale. She certainly had trouble initiating conversation and handling change.

I read that early on, the FTD patient's memory tends to remain intact, whereas in Alzheimer's disease, short-term memory loss is usually one of the first symptoms. Okay, I thought, Ma had certainly had memory problems, but these had been clouded by the effects of drinking. Hadn't the changes in her personality been more noticeable, and hadn't they started much earlier? The apathy, her waning social skills that had saddened me at our wedding? (She had been sixty-five at our wedding, precisely the average age of onset of FTD.) And now the disinhibition. All these behaviors were associated with frontal lobe damage. She still wasn't having any major language issues besides the occasional struggle to find a word—and I suffered from that myself. But couldn't this actually be the frontal form of FTD? Pieces fell into place: I had never been convinced that the Aricept had helped Ma's memory, although it seemed to make her more excitable. Now I read that Aricept and similar drugs had proved ineffective, and in some cases even counterproductive, in FTD patients.

When I finally stopped reading to glance at the clock, I panicked; I had to supervise a lab in ten minutes. I raced back to unlock the door for my students, reluctantly putting off my investigation, but the following afternoon I holed up in my office with the door closed and called the Alzheimer's clinic to press the doctors about what I had learned.

"You've really done your homework," Nurse Carol said. I

could hear her leafing through Ma's chart as she spoke. "It looks like the doctors have not completely ruled out frontal lobe dementia in Ruthie's case."

The CT scans, she said, showed Ma's frontal lobes had been more shrunken than other parts of her cortex, which was consistent with the frontotemporal dementias. However, her memory problems—the little notes to herself, the trouble keeping track of recent events—were more common in Alzheimer's patients. I was scribbling notes on the back of the letter they'd sent us after her diagnosis.

"Wow. I really wonder which it is. But it seems like if we wait, it'll become more clear, right?" A dangling factoid was nagging at me. I had read that the FTD class of dementias had a much shorter time course than Alzheimer's disease; whereas Alzheimer's patients could linger for decades, FTD took its subjects in an average of three years.

"Well, yes, that tends to be the case. But you know, Sybil, this kind of mixed diagnosis is fairly common. I wouldn't worry about it too much. Unfortunately, in either case what it means for you is the same: your mom will need care, more care as she goes along."

"Right, of course. I'm just trying to understand all this, you know?"

"Of course you are." She paused and asked after Ma. Was she still managing okay at home? And me, had I called the caregivers' support group she'd recommended?

"Um, no. Not yet. But I have the number. It's just, those groups seem to all meet at the most impossible times."

She recommended that I try anyway. Her last words were "You don't have to go through this alone."

I didn't tell her that I didn't feel comfortable leaving Zoë with Pat to attend an evening support group, because I was afraid I'd already driven him close to his limit. So much of my energy was taken up by the pregnancy, and so much of our free time by Ma's visits.

Pat knew he'd do the same for his own mother. He knew this was the right thing, the decent thing, to do, and he was a decent man. Plus, despite all the hassle, he loved my mom. So he stopped short of asking me not to invite her, just as he stopped short of telling me he wished I hadn't gotten pregnant so soon, but I could feel the tension in him. Pat had softened some as he watched my belly grow, but his emotional support was nowhere near what it had been during my first pregnancy. That sweet you-and-me-against-the-world feeling we had shared as we holed up in our tiny Waltham apartment in the snowstorm to await Zoë's arrival was gone, and in its place was a growing remove—he continued to return home late, sat apart, and played video games late into the night. Pressured by the added responsibility of a new baby on the way, he commuted his hour and a half each way to work and regarded my upcoming maternity leave and summer off with open envy. He didn't come out and say it right away, but he seemed to feel that I was getting what I needed and he was not. And fearing he might be right, I didn't feel like I deserved additional support. It felt as if I had already asked for far too much.

I didn't tell all this to Carol. When I got off the phone, I didn't call the support group number. I went right back to the question of Ma. I adjusted the picture I had constructed of her brain, taking into consideration that FTD was a real possibility. If we could see inside without cutting it open, we might not find the amyloid plaques and tangles of tau protein that I had envisioned the day we received the tentative diagnosis of Alzheimer's. The damage might not be primarily buried beneath her temple in the curling folds of her hippocampus after all. If it was FTD, we'd look instead to her frontal cortex. There we would see white matter lesions—degradation of the sheaths that insulated her axons. Her neurons would be riddled with mysterious little holes called microvacuoles, tiny empty spaces opening up in their cell bodies, turning gray matter to Swiss cheese as the neurons died away. Into these gaps large glial cells (the nonneuronal cells of the brain) would extend their creeping tendrils to form a dense fibrous complex, patching together the remaining neurons in a network of glial scars.

But we couldn't see inside. Ma wasn't a tree; we couldn't take a core sample. The behaviors associated with the different dementias overlapped, and we couldn't know precisely what had happened. I was starting to better understand why her doctors hadn't been able to give a definitive answer without an autopsy.

Zoë's three-year-old prefrontal cortex just happened to be going through its own period of rapid change right about then. Following an intense developmental growth period, the density of her prefrontal neurons was now on the decrease, too. But whereas Ma's prefrontal gray matter had already thinned to *below* normal levels, Zoë's was still dropping down *to* normal levels. Synapses that had formed in utero and during the first year or so after she was born were now being pruned away, as relevant connections grew stronger and irrelevant ones atrophied.

As this happened, her ability to perform certain memory tasks was improving dramatically. These were tasks that required her to remember a rule while simultaneously suppressing a response that came naturally, the way one must do in a game of Simon Says. I watched Zoë and Ma at the dining room table after dinner one night, each of them trying to remember to pat her head only if Simon said so, each of them attempting to inhibit the impulse to obey my command. I was happy because for once Ma was fully engaged; she was playing with us.

"Simon says stand up!" I cried. Towheaded Zoë and gray-haired Gram both pushed their chairs back to stand up, wearing identical wild, expectant grins, their cheeks flushed with excitement.

"Simon says touch your nose! Simon says touch your toes! . . . Now, touch your nose again!"

"Ha! She got you!" Ma exclaimed triumphantly when Zoë finally slipped up.

"Again, again!" Zoë squealed, up for the challenge, but Ma didn't want to play anymore. We were all aware that Zoë had been gaining on her; she wanted to quit while she was ahead.

Ma and Zoë were both in a state of flux as far as rules and inhibition were concerned. As Zoë gained the ability to inhibit certain responses, Ma was losing her ability to inhibit others. I wondered if there would be a moment when they would precisely balance each other, when they would be prefrontal equals. I wondered if that moment had already passed.

The following Friday afternoon Zoë asked Ma to read her a story. They sat cozily on the couch together, Zoë looking on as Ma read in her teacher voice from a Winnie-the-Pooh book. I was eavesdropping, hungry for this moment of apparent normality in my mom's role as a grandma.

But Zoë, in typical exuberant preschooler fashion, kept jumping in eagerly: "Look, Tigger's tail is a SPRING!"

The first time Zoë interrupted, Ma nodded indulgently. The second time she said, "Yes, okay, but let's continue with the story now."

Zoë, irrepressible, kept it up. "Gram! Watch me do the Tigger Bounce!"

She was bouncing on the couch now, with one hand on Ma's back, jostling the book up and down.

Suddenly Ma pulled her arm back and threw the book to the floor, hard. "You're RUINING the STORY!"

Zoë stopped bouncing, surprised, eyes wide and indignant. "I am not!"

I felt my face go hot. Ma threw a book! Ever since her explosion at Saul's, I had been watching Ma and wondering what I would do if Zoë found herself on the receiving end of Ma's unchecked anger. *Here we go,* I thought.

"You are TOO!" Ma was insisting.

"Am not!"

"God DAMN it, I'm not reading to you ANYMORE."

"I don't care, and no throwing books!"

As Zoë stomped angrily away, my heart was racing—but just as quickly, it was sinking. In the wake of her swearing and her anger, my vision of Ma as the patient, wise old grandma for my Zoë was fading fast. Zoë had just learned that Ma could not be the grown-up with her. She was quoting Ma the rules. I was afraid she'd begin to see Ma as weak and inconsistent, that she would begin to distrust her, that they would never grow close; I was afraid that it was too late for them to really know each other.

I never did bond with my own grandmother. The last time I made the trip to Laguna Beach to see Grandma, I was twelve years old, and everything had changed. Gramps was dead, Grandma was in her eighties, and my sister was on a special school trip and didn't come along. I was too young to go to the beach alone, and too old not to notice the stale stillness of their house. It smelled of puffed wheat and sour soil. The heavy drapes stayed closed all day. I tiptoed through the hushed rooms, recalling our voices. I slinked into my dead grandpa's private no-frills bathroom, where I used to preshower after a day in the ocean so as not to get sand in Grandma's pink bathtub. Peeking across the dark hall, I listened to the voices of my mother and hers in Grandma's lavatory.

"Okay, Mom, you ready? Let's go—one, two, and three . . ."

The door was open a crack, and I caught a glimpse of Ma's flushed face as her strong, freckled arms grasped Grandma's skinny, pale ones. As Ma lowered her painfully down into the warm bath, Grandma gasped. Thin, faded skin flapped uselessly over her bones where muscles should be. Grandma hated needing assistance from Ma or anyone else. Helpless, fragile, she cringed there in the water, and I heard her small cry of dismay, of shame.

At that moment my empathy for Grandma slipped. We had never been that close to begin with, and now her flaccid limbs, her powerlessness, revolted me. Although I knew it was wrong and cruel, I couldn't help myself; I began to distance myself from her, to see her as foreign, untouchable.

I hated the thought of Zoë feeling this way about her Gram—indifferent, cold, and distrustful. I hated the idea of Ma dying without having known the love of my joyful little scamp of a daughter.

Now I stood between them, uncertain what to do. I could enter the living room, pick up the rumpled book, and talk to Ma about her outburst, or I could follow Zoë into the dining room, try to comfort her, and ask her how she felt about Gram. Was she afraid? I turned and started toward the dining room, only to bump into Zoë, who was heading back to the living room; Zoë, who was not twelve-year-old me, who was not a squeamish, judgmental preteen but an enthusiastic, openhearted toddler, who lived in the moment, and who was more fond of Ma than I had ever been of my grandmother. She padded past me holding a big plastic box and asked matter-of-factly, "Wanna play Legos, Gram?"

I turned away and stared out the living room window, searching out a distant little blue patch of bay through the trees. I was smiling, even as the tears came. This was the gift Zoë brought to my mom and me: she didn't have any expectation of how a grandma *should* be, so she gladly accepted this one. She was just happy to have another person in the house to play with, and like most three-and-a-half-year-olds, she was resilient; she was forgiving. Ma didn't usually bring presents anymore, she didn't cook meatballs or do complicated art projects like Pat's mom did, and yes, she had even yelled and thrown a book, but Zoë didn't mind all that—only I did. These disinhibition incidents would continue in the months to come. I was careful not to leave the two of them alone together after that day, but Zoë seemed to take it all in stride. Gram was just Gram, and so far Zoë still loved her plenty.

CHAPTER 13

Baggage

Zoë was napping; we had gone to the park after breakfast, and she had conked out in the car on the way back. Ma had returned to her house in Martinez, and after putting Zoë to bed, I was reassembling the futon into a couch. Ma had neatly folded her blankets and left them, not behind the couch where I kept them, but piled on an armchair on the other side of the room. No matter how many times I told her, she couldn't get this right. Was that yet another symptom, or just a quirk of her personality? On top of the blanket pile lay the thick flannel pajamas I kept here for her. As I lifted the stack, I eyed the crotch of the pants: the otherwise smooth baby-blue fabric was damp, yellow, and wrinkled there. I put down the blankets and held up the pajamas, sucking in my breath.

A thing as simple as this. I knew that many Alzheimer's patients began to experience incontinence in the middle stages of their illness. Accidents would happen with increasing frequency over a period of years, until the patient completely lost bladder control. Eventually control over the bowels would also go, but that usually came later. If these had been Zoë's pajamas, I would have matter-of-factly plopped them into the clothes hamper with a smile and a sigh. I'd have reassured her that her bladder just had to grow and learn, that this was perfectly normal. But that patch of urine-stained fabric sent me in such a different direction, because it was my mother's.

I felt I had a direct tap into her dismal future. She wouldn't just get old and weak as Grandma had. I remembered Christmas-caroling one year on the Alzheimer's ward of a local nursing home, that group of silent, white-haired women in the lobby, slumped in their wheelchairs, sleeping or just staring. End-stage. I took it further, envisioning a room with dim fluorescent lights, chrome bed, blinking machines, liquid slowly dripping from motionless hanging bags into multiple tubes. In the bed lay a failing body, a life no longer lived but only suffered.

For the rest of the day after finding the pajamas, I was slightly irritable. Two a.m. found me sobbing into my pillow beside a sleeping Pat. I didn't want to wake him, but I desperately wanted someone to take care of me. I needed a friend to tell me everything would be okay. I tiptoed into Zoë's room and breathed in the sweet-salty scent of her sweaty head in great, needy gulps.

I wondered obsessively whether Ma had felt that same desperate need to smell me, to touch me, back when I was too young to remember. She must have. But when I questioned her about her mothering (Did you watch me while I slept? Did you carry me in a sling?), the answer was always the same: "Oh, I'm really not sure. I suppose probably so." She seemed so uninterested. I wasn't sure whether she actually couldn't recall or whether these things simply no longer mattered to her. Either way I found it distressing. I was on a mission, to find out all I could about Ma as a mother and absorb it, before it was too late. I felt compelled to guard whatever memory of those days she did retain, to keep them fresh for her and for me. In the way I mothered Zoë, I continuously reminded Ma and myself of how she had mothered me.

I had felt this one day in the kitchen as I stood behind Zoë and she rolled pie dough with our big marble rolling pin, with Ma looking on.

"You're such a good pie maker," Ma remarked as I emptied our freshly picked blackberries into the shell.

"I should be," I said with a smile. "I learned from the very best pie maker I know."

"You mean me!"

"Yes, I mean you. And now it's Zoë's turn, isn't it, babe?"

Zoë smiled up at me as she dotted the berries with little blobs of butter. "Is this good?"

"It's better than I could do," Ma said.

She was right, and she was wrong, depending on what time frame we considered. I watched Ma's face as she stepped forward, standing close behind Zoë's and observing with interest. Why couldn't she make a pie anymore? She was still in the early phases of her disease, with her short-term memory beginning to wane. She should still retain at least the basic memory of pie making, I thought, since she had learned it as a child herself and long-term memories usually remain intact much longer.

She was still watching Zoë dot the butter. "I think you need some on the other side, darling," she said. "Over there."

In fact, she did seem to remember in a general way. Maybe she simply could no longer pull it off herself; maybe there were too many elements to manage at once, and too much physical coordination involved.

I liked that she called Zoë "darling." That was what she had always called me, darling, punkin, sweetheart. Now I called Zoë punkin, sweetie, babe. When I used punkin in particular, I was deliberately evoking Ma. Sometimes consciously and sometimes unconsciously, I had begun to emulate my mother, the young Ma, the one I had once called Mommy. At times I felt it so potently, as if I were time-traveling back to 1965; as I knelt before Zoë smearing Coppertone sun lotion on her sweet pink cheeks, calling her Punkin, I thought I might actually become Ma for a moment, the overlap was so complete. As Zoë and I snuggled on the couch for a good read, Mommy and little Sybil were there, too; I was nestled up against her warm, soft body feeling safe and loved. I heard her voice in mine as I read the very same stories: *Ferdinand,* Hansel and Gretel, *Green Eggs and Ham.* In this way I melded with her, I became her, even as she became not-her.

The emulation did not extend to all corners of our lives, however. There was one major way in which I differed from Ma as a mother, and it had particular bearing in the context of her gradual fading. Throughout my childhood, every Friday night without fail, Ma drank, and more often than not, she got drunk. My father did, too. This had been part of their life in college and part of mine from birth, a fact I just accepted as the norm in my childhood. Their Friday night drinking was integral to our family social life. My parents' friends showed up around five p.m., and a troop of kids romped about madly while the grown-ups talked and laughed and drank, sang, and played guitar. Around seven we'd all sit down for chicken and rice, and sometime after dinner the other families would drive merrily off into the night, and we would go to bed.

By the time I'd begun to develop breasts, though, the Friday night ritual had started to change. The other families had stopped showing up. The drinking began earlier in the evening and ended later. It began earlier in the week, too, and continued through Sunday night. The year I entered eighth grade, sixteen-year-old Alice left, with my father's blessings, on a six-month tour of the United States with her older lover. That summer, free from work and distressed at Alice's departure, Ma took to drinking every night.

This was the Ruthie most people never saw, the private Ruthie. Not the schoolteacher who returned home from work full of concern for her first-grade students, not the peace activist, not the wholesome, thoughtful neighbor who delivered bags of oranges and tomatoes from her garden to old Mrs. Fraga next door; this was just a sad woman drinking. Before dinner she'd have a few Manhattans, and with dinner she'd drink beer. As I cleared the dishes from the table, she would open one more, and a little while later I would see her eyelids start to droop, her head begin to nod.

A helpless, desperate sadness would well up in me. Many times I tried to get her to go to sleep before it happened. "Mommy," I pleaded, "just go to bed." I didn't know enough then to call what she did "passing out." I couldn't put my finger on why I hated seeing her "asleep" at the table; it just wasn't how I wanted her to be. In retrospect, it's not surprising that I overlooked the early symptoms of Alzheimer's disease with such ease; for so many years I had been forced to accept this other form of intermittent absence and confusion.

One hot summer evening after dinner had been cleared from the table and my parents had already had a few, Ma was complaining about the heat.

"Why don't you just take off your top?" Daddy said. Ma blushed and shook her head.

"Why not? It's just us. That could be pretty sexy, you drinking your beer topless."

Thirteen years old and not particularly precocious, I squirmed in my seat. Daddy had always flirted openly with Ma in front of us kids. This had always seemed harmless and silly to me, and it probably was, but recently I'd become less comfortable with it, along with all other mention of sex. At least I knew Ma would never do it.

And then she did. She just reached down and pulled off her blouse. I stared, incredulous. Her fleshy middle puddled below her white cotton bra, and large, pale, cone-shaped breasts pushed out over the tops of the cups. Her jowls were loose, her eyes watery and defiant as she reached for her beer and slugged down half of it.

"There," she said, setting the mug down too hard on the Formica table. "How do you like that?" A little ridge of foam clung to her upper lip.

I groaned, "Oh God, Ma . . ." Drinking her beer in her white cotton bra. Had she been sober and in slightly better shape, her taking off her blouse might have struck me as audacious—I was old enough to appreciate Ma as the spirited, modern woman she was to her friends and co-workers. A few years earlier she had led her fellow teachers in a strike to change the dress code at her school and had won women the right to wear pants to work. Whenever her best friend, Vicky, came by, she made sure to remind me and Alice of what our mom had done for the teachers of Pittsburg Elementary School. Taking off her blouse could be seen as an act of defiance, a spirited rejection of the rules that bound her. That summer night, though, Ma was not spirited, she was pitiful. Inexplicably frozen in place, I sat stiffly at the table across from her and watched as she finished that beer and started another before her chin slowly sank to her chest, her eyes drooping closed. I would come to resent the fact that this particular memory was so permanently imprinted in my cortex.

Pat had told me that when his grandma Grackle got dementia, everyone assumed it was caused by alcohol. Ever since then I had assumed that this was a known phenomenon, but it turns out that there is a fair amount of disagreement as to whether alcohol-induced dementia is an illness unto itself. Few doubt that alcoholism exacerbates dementia, but exactly how the two conditions are related remains unclear.

One well-known effect of heavy alcohol use is that it interferes with the breakdown of thiamine (vitamin B1) in the body. Even with a well-balanced diet, if a person consumes enough alcohol, chances are her thiamine will not be absorbed. This thiamine deficiency causes two very well-documented neurological dysfunctions in advanced alcoholics.

The first, Wernicke's encephalopathy, is a severe, acute, and life-threatening disorder that causes patients to become confused and disoriented, lose short-term memory, and suffer damage to the nerves that control their eye movements and coordinate their lower extremities. They stumble and stare; they are bewildered and overwhelmed. This condition is often followed by the second, equally serious illness, which is called Korsakoff's psychosis. Korsakoff's patients, like midstage to advanced Alzheimer's patients, typically have severe anterograde amnesia—trouble forming new memories. A Korsakoff's patient might give a detailed account of the events of her childhood over breakfast but by lunchtime have no recollection of either the conversation or the breakfast. For some reason it is very common for Korsakoff's patients to confabulate (give false accounts of the forgotten events, believing they are true) to compensate for the missing memories.

Ma had known decades earlier that alcohol depletes vitamin B; for years she had religiously washed down a vitamin B complex with her grapefruit juice on the mornings after her nights of heavy drinking. She told me she had read that if you drank, you had to do this in order to stay sharp. I remembered feeling glad that she was looking after herself so well. And maybe it had helped. These symptoms sounded much more extreme and physical than Ma's.

Her recent memory was sometimes foggy, but clearly she wasn't suffering from any of the motor symptoms of Wernicke's encephalopathy, and she did not confabulate.

Wernicke's and Korsakoff's are not caused by direct effects of the alcohol itself—they are a result of concurrent malnutrition and alcohol-induced thiamine deficiency. But newer studies imply that alcohol may have contributed to Ma's troubles in other ways. Recent research demonstrates the presence of a slough of physiological problems in the brain—especially the frontal lobes—of alcoholics.

One effect that could be considered almost poetic concerns the cell membrane. Neurons and glial cells, like all the cells of the body, are surrounded by this membrane. It consists of a double layer of phospholipids—greasy molecules that are not soluble in water and that thus serve as a protective outer boundary that controls what substances enter and exit the cell. Long-term alcohol exposure, however, creates a problem for the cell membrane. With prolonged exposure, it has trouble, not when the alcohol is applied, but when it is removed. When the alcohol is taken away, the membrane, which is normally soft and malleable, becomes rigid and unyielding and will soften again only when returned into the presence of alcohol. This struck me as such a perfect analogy for an alcoholic person's behavior, I had to laugh out loud when I read it; however, it also sounded very serious. Proteins lodged in cell membranes perform countless functions within and between cells. If the membrane becomes rigid, these proteins will not be able to move as freely within it, which could prove disruptive to any number of cellular processes.

Other problems arise in the brains of chronic alcoholics. For example, they suffer from white matter lesions—a wounding of the material that insulates nerves in order to speed conduction of nerve impulses. In the prefrontal cortex of the alcoholic, metabolism tends to slow down, and blood flow is poor; glial and neuronal cells are more sparse, and their nuclei (the small sacs where the DNA is stored) appear shrunken. I wondered whether the impact of alcohol in the prefrontal cortex predisposed a person to get frontotemporal dementia or Alzheimer's, or the other way around. I could imagine a scenario in which the dementia started much earlier than we knew and combined with a genetic predisposition toward alcoholism, setting up a tendency toward disinhibited alcoholic drinking. This in turn might lead to frontal lobe damage, and thus to further disinhibition and maybe further alcoholic drinking. Although I found no research to support or refute this idea, I was struck by how easily Alzheimer's disease, FTD, and the effects of long-term alcoholic drinking could be confused, especially early on.

Precisely how all these alcohol induced changes affected behavior was not clear to me, but shrinkage, lesions, slowed-down cells, and rigid membranes obviously didn't bode well. I began to associate these cellular degradations with the image that had never left me, of Ma slumped in her chair, sweaty and beery and still half naked as she began to snore. Maybe the strength of that image explained the unrelenting, self-righteous fury that sometimes boiled up in me when Ma slouched wordlessly at my table, or couldn't remember a conversation we'd had the previous evening. Maybe if she hadn't been so drunk so often, her supply of brain cells would have lasted longer.

The morning she was finally diagnosed, doctors in white lab coats with stethoscopes and clipboards had told Ma plainly that alcohol exacerbated her dementia and that she should cease all drinking immediately. Only then had she finally stopped. *Great,* I had thought. *Now that it's too late.* It may not have been fair of me, but in the deepest, hardest way, I secretly blamed my sick mother for the time we had lost.

•   •   •

When we sat down to eat our pie, Ma sighed with satisfaction. I wondered how it was that she could lose enthusiasm for just about every other thing in life while maintaining this undying affection for food.

"You really do know how to make a pie," she said.

"So I hear," I replied with a smile. "But you know, I had the very best teacher."

"Huh." Ma didn't look up, only nodded absently as she held out her plate for seconds.

CHAPTER 14

Number Two

I was caught between the physical world and the abstract swirl of a deep sleep. Sunshine streamed through the lace curtains into our bedroom, and birds chattered in the yard, but I didn't open my eyes. My body was a beached whale, a fleshy mountain, still and solid and immovable, drenched in exhaustion. At first I was not thinking or feeling anything, but something nagged at me; something major had shifted. The way a person just home from a sea voyage still feels the pitch of the waves even on dry land, for a moment I was sure the baby was stirring within me, but as my mind woke up, I was aware that this couldn't be, because . . . and then I caught it: the sound of a tiny breath in my left ear. Baby! It was April 2001, and our second baby girl had just arrived.

I opened my eyes and turned as the happy realization spread through me. Cleo, four days old, pink and round, lay on her back between Pat and me, her arms flung straight out at right angles to her body, her little chest rising and falling under the terry-cloth onesie that was already almost too tight. She was so much bigger than Zoë had been as an infant. *Born big,* I thought. I looked up.

Pat was on his side, smiling at me, smiling at Cleo. If I'd been worried that his resentment over this pregnancy would affect him as a father, all the worries had left me as soon as she was born. He was so clearly overwhelmed by his love for this new baby; she was his, and he adored her without reservation.

"Good morning," he whispered. His long hair lay in a tangle across the pillow. His chin was dark with stubble, his eyes puffy with sleep.

"Heyyy," I said, "for a second I forgot—I thought she was still inside me."

"I'm surprised you're even awake right now."

I smiled. "I know, I think she was hungry." Pat had stayed up watching a movie while I had drifted in and out of sleep, but I must have nursed her at least six times. I tried to prop myself up. "Ow. Oh, damn. This again."

It had not been an easy birth. Everyone had assured me the second one would go smoothly, but this had proved oh so wrong. We'd had no midwives this time; foolishly we had thought our vast experience with childbirth would get us through. In the middle of the night we'd had medical intervention after medical intervention, and after twelve hours of hard back labor, I had finally undergone an emergency C-section. My somewhat uninterested male OB hadn't shown up until the next day, and then he had told me I looked fine and attempted to dose me with Percocet.

In one respect and one only, the trauma had been a good thing: Pat had really rallied for me. I had heard the defiance in his voice when they tried to administer drugs we had not asked for, seen the fear in his eyes when the nurse hit the wall button and yelled, "We need some help in here!" I'd felt his love and anxiety and dedication as he watched them cut me, and his relief and instantaneous fatherly devotion as soon as Cleo had slipped out with a husky yowl.

After she was safely out, when the doctor had removed my entire uterus and lain it out on my belly to sew it back up, Pat had watched the whole thing.

"What is that?" he asked.

"Well, that's her uterus." The anesthesiologist was back with us, behind the paper partition, watching with Pat and monitoring my meds.

"Wow," Pat said, "that's amazing!" He looked down at me, squeezing my hand. "You want me to describe it to you?"

"Um, maybe later," I mumbled weakly. "Hey, aren't I supposed to be the biologist in this family?" But I was proud of him, and I was relieved. For the first time in months we felt like a team. It was us against doctors, us against death. Pat didn't want to lose me or Cleo. It was good to know.

Now I was not allowed to climb stairs for a week, to lift anything for two, or to lift anything heavier than Cleo for three more after that. Every morning I had to be reminded again: no, you can't go make breakfast, no, you can't clean the house no matter how disgusting it gets, or make the dinner or do the shopping; you have to let Pat or someone else do it. And now here came Zoë, padding over to my side of the bed with her arms outstretched. I was in such trouble.

"Sweetie!" I reached out for her but then stopped. "Can you climb up here by yourself? And remember, I have a big owie, so you can get next to me but not on top, okay?"

Just as Zoë snuggled in with me, the phone rang. Pat got it, put on his work voice. "This is Patrick." He looked over at me, his voice softening. "Hi, Ruthie. No, we're up. Yes? Yes, everything's good here. And how are you doing?"

I could hear the pitch and rhythm of my mother's telephone-compressed voice but not the individual words. I felt a weight in my chest; it was Thursday. Normally I would have called her by now. In fact, I had been not too guiltily basking in the luxury of having an excuse not to call, not to invite her for the weekend. I couldn't wait for Pat to take Zoë to preschool so I could snuggle back into the bed and nurse.

Pat handed me the receiver, which I took with a reluctant sigh. "Hey, Ma, how are you doing?"

She asked how I was feeling and about the birth. It was sweet, I thought, that she was making such an effort. I started to tell her all about the amazing baby, and Zoë squirmed on the bed.

"And you should see the big sister—she loves that baby so much! Don't you, sweetie, you love your little baby?" I had been making a big effort to include Zoë, to give her a sense of entitlement. Since I'd gone into labor on Zoë's fourth birthday, I was encouraging her to see Cleo as a gift to her, not just a horrible intrusion on her perfect parental-love monopoly. Zoë smiled up at me and planted a tiny, gentle kiss on Cleo's temple. So far it was working.

"Well, darling, I'll let you get back to the baby. I know I probably won't see you this week, and that's fine—"

"Ma, I'd love for you to come out and see the baby—"

"No, now, Daniel said you might need some time, and I understand that. I think I'll just come by next week. I can take care of myself—I always have."

I breathed a sigh of relief. A part of me cringed at the thought of her alone for a whole week, but I also just wanted a few days of adjustment, and for the moment I was off the hook; *thank you, Daniel.* I told her I would call soon and hung up. She seemed fine, very upbeat and self-assured. Several weeks later, I would find a note on her dining room table: "Call Sybil: Ask about BABY. Name = CLEO. Friday night—NO. Don't ask!"

Pat took Zoë downstairs for breakfast. Cleo was smacking her baby lips and making throaty little calls. Her voice was deep for an infant's, and this together with her size gave the impression of an older baby. She was also a voracious and loud eater; with every gulp she made a loud and satisfied little grunt, *uhg-ahh,* and I had started calling her Monica Seles, after the tennis player known for the sensational power grunts she emitted every time she hit the ball. I liked Cleo's deep voice, her crazy appetite, and her heft; I liked her. And because my maternity leave would overlap with the end of the semester, I was going to get over three full months of snuggle time with her. As I settled her at my breast, I could feel myself relaxing into the downtime in a most delicious way.

On my last day at work my students had had a long, difficult lab, and we had all worked hard, but at the same time it had felt like a party. It was the end of a big week, and we were all a bit punchy. It was also a preschool holiday, so not only was I inordinately large and somewhat of a celebrity for the day, but I also had Zoë with me in my office. Students fawned over her, asking her how she felt about becoming a big sister, and she hid behind me, smiling and blushing and whispering her replies into the back of my leg. Everyone wanted to know if the baby was a boy or a girl, how I felt, whether we had chosen a name; they brought me cards and gifts. One boy who worked in the botanical garden on the hill came by with seeds for my garden, exotic red carrots and African wildflowers. I pictured myself lovingly conditioning the soil and planting wee seedlings while Zoë played in the wading pool and the baby slept in a basket in the shade.

Poor Pat was in another world entirely. His job in San Francisco had imploded in the midst of the dot-com bust two months earlier, but he had made an impressive recovery, taking only two weeks to secure a new position at another start-up much closer to home. Watching him navigate the business world with so much finesse, I had a moment of admiration, and also a tiny ray of insight into our differences: he was in the middle of a steep career trajectory, I realized, learning everything he could as fast as he could and excelling at all of it. I reminded myself that he was six years younger. Maybe if he'd married someone closer to his own age, he could have waited for a less insane moment in his professional life to start a family.

This new job held a lot of promise for our family; gone was his long commute, and I had fantasies of meeting him for lunch and of having regular family dinners. Unfortunately, so far he had been working so many more hours as he came up to speed that if anything, Zoë and I had seen less of him. In fact, I'd been seeing less of everyone. I had turned into a pregnant recluse, like a great blimp floating above the city of Berkeley, just waiting.

I hadn't been able to knock the feeling that I was playing hooky. I'd drift from room to room in our house, switch on the TV out of idle curiosity, and then, feeling myself drawn into the vortex, switch it off again in horror: dreary hospital dramas, hemorrhoid medicine and diet pills, Huggies diapers and household cleaners! I was desperately afraid of becoming the person (let's face it, the woman, the housewife) whom those shows and ads targeted. When I drove downtown at midmorning, it seemed to me that the whole world was at work. I was like Vincent Price in the movie *The Last Man on Earth:* aimless, useless, and spooked.

I avoided the neighborhood adjacent to the university, afraid I'd encounter my co-workers or students on their lunch break. What if they caught me ambling widely into the hardware store in the middle of the afternoon, when I could have been teaching an intensive lab course to a group of eager undergrads? I knew precisely where I'd be in my workday at any given moment, and I recalled exactly the acoustics of my long lab room, the hum of machinery, and the smell of cockroaches, formaldehyde, or computer-generated ozone, depending on the lesson of the day. My head was still full of bits of information: details of dissections, recipes for solutions, neuroanatomy factoids. I worried about the assistant I had hired to replace me through the end of the semester, talked too long when he called for help, and in my journal sketched out possible improvements on the lab section the class was completing in my absence.

Now that the baby was born, I was still thinking about neurobiology, but my mind was no longer in the lab; just as when Zoë had arrived, my thoughts on biology had become much more personal. My subjects were now Cleo, Zoë, and Ma, and they provided me with a nice spread of developmental stages.

Neurobiologically, Cleo was still in the Play-Doh stage of life. Her body had produced tens of billions of brain cells while she was in utero, and ever since they had come to be, these neurons had been learning to communicate with one another, reaching out axons and forming synapses. Every experience she had was changing the architecture of those connections in her brain. As she lay beside me on the bed softly gurgling, she was at the start of that developmental period in which kids can learn any language from Portuguese to Punjabi with equal agility. Yet as early as six months into her life, she would begin to specialize, to attend specifically to types of sound unique to the English language. Before her first birthday, she would be able to discern the structural units of English, and she would become sensitive to word order even before she could speak.

At four, Zoë was only slightly less mentally malleable. She was no longer a linguistic "citizen of the world" with equal potential in all languages like her sister, but she had recently acquired an ability that Cleo still lacked: deep in her brain, cells nestled in the folds of her hippocampus had begun to store her first conscious long-term memories. And as if to confirm the fact that what goes up must come down, Ma's memories had begun to float away in a flutter of time, like dandelion seeds in a summer breeze.

The morning of Ma's first visit to the house after Cleo was born, Pat suggested that I might enlist her to do some simple cleanup around the house; I was bedridden, after all, and she was perfectly able-bodied, he reminded me before heading off to work. This seemed reasonable to me. I asked her to fold laundry.

"Sure I can," she said a little too loudly. "Anything you need, darling."

Folding laundry always took me back to my childhood. I could picture my parents' bed piled with fresh clothes just pulled from the outdoor line that ran from our back porch to a pole halfway down the yard: my ripped and patched jeans, my dad's green work pants, our tie-dyed T-shirts, and the colorful vests Ma made for herself and Daddy on her Singer sewing machine. I remembered sitting beside her, watching and imitating as she laid a blouse on the bed facedown and tucked back the sleeves before flipping the bottom half up longways. When she turned it over, it looked just like a shirt on the shelf at Hinks department store downtown. I flipped mine, and it looked like a rumpled, crooked version of hers.

Now Ma was sitting at the end of my bed with the laundry basket at her feet and one of Pat's black T-shirts in her lap. "Now, let's see . . . do you do it like this?" she asked, loosely folding the shirt in half and then in half again, like a sheet of paper.

At first I thought she was inquiring as to my particular style of folding. I looked up at her face. Her eyebrows furrowed as she concentrated, her mouth frozen into a frown. She unfolded the shirt and then tried again. It reminded me of the way I pick out a song on the guitar after months of not playing, willing the motor memory to kick in because it's an unconscious process and I know it will come if I just begin to play. But it didn't come for Ma. Then I realized that what she wanted to know was not how I, Sybil, folded clothes, but how clothes folding was done generally.

Hey, not fair, I wanted to say. Procedural memory was one form of memory that was normally spared in early Alzheimer's disease; it relied on different brain structures than did memories of events, facts, and personal experiences. Remembering how to fold laundry should be more automatic than, say, remembering a neighbor's name. But this was one more reminder that everything I knew about Alzheimer's was statistical, and statistics, which I had always believed in so vehemently, absolutely lost their strength now that I was dealing with a single real-life example. Ma wasn't necessarily the average dementia patient. The disease might be advancing faster than I had supposed, or she might just have a slightly different distribution of beta-amyloid in her brain tissue, and there was no controlling it; just as with anything else, the best I could do was to stay in the moment and be kind to her.

I tried to imagine how it must feel, to lose an ability as basic as laundry folding. She must have understood that this was something she *should* be able to do. Perhaps she had started in with confidence, only to find the task bewildering and elusive, the feel of the fabric foreign in her hands.

"Yeah, that's fine," I finally said. "Here, hand me a few, and I'll do them with you."

After Ma left that afternoon, I sat and refolded the laundry from her pile. Someone was hammering in the yard next door, and a squirrel scolded a mockingbird from the back fence. I smiled down at Cleo, who lay sacked out with a just-nursed blush in her sweet round cheeks, surrounded by odd socks and dish towels in the middle of the bed. Even with the C-section trauma, everything about this second baby seemed to be easier; she nursed more easily, she slept more easily, and I took it more easily. I knew how to do everything and what not to worry about as far as Cleo was concerned. As to Ma, I didn't know anything. I didn't know anything at all.

CHAPTER 15

Regrouping

In the ensuing days, nursing, sleep deprivation (that absolute biological brick wall), and the mandated bed rest all conspired to slow me down to a pace I'd never experienced—not even when Zoë was born. It was as though a great pair of hands had reached into my life and grasped the two ends of the time line, stretching it gently but persistently. The longer I stayed in bed with the baby at my breast, the further the moments extended outward, the line breaking into its many points, the space between the points expanding, until I sat suspended in time, coasting almost imperceptibly along.

I did something I didn't fully understand but that felt essential: for two solid weeks after the day we'd folded laundry together, I didn't invite Ma back. Then for the next two months I invited her only every other week and just for Friday nights. I knew this left Ma lonely and put an extra burden on Daniel, who lived so much nearer to her than I did. I knew she was needing me more, not less, and I did call her, but without even perceiving it as a choice, I retreated to the nest; I wanted to see only Pat and my babes.

As my body healed and I began to venture out into the world, the peculiar time warp deepened, opened, and swallowed me. I sat in a daze at the toddler park with Cleo on a blanket in the grass while Zoë poured sand from bucket to bucket. It was a low-key routine. At midday the park was quiet, the energy low. Toddlers wobbled about, and every once in a while one fell over. Moms, nannies, and the odd dad watched idly, followed along patiently, and stepped in and caught them just before they waddled into a disaster.

Parenthood (and nannyhood) seemed to make the grown-ups uniform and bland. We submerged ourselves in this life—a sort of dirty pink color, I imagined it—and we all got coated in it, by the common themes: encouraging, comforting, nurturing, scolding, teaching, protecting, diapering, and inspiring our young charges. We professionals allowed our career passions to take the backseat for the time being; our work personalities faded a little, and the day-to-day issues that had been so pressingly important the month before seemed much less so during the days we spent suspended in our muted reality. It was as though the frequency range of our selves had been compressed to the point of distortion; all our colors, shapes, sizes, religions, sexual orientations, and vocations had been temporarily eliminated, and all that remained was parenthood. We were inarticulate and languid, our middles soft, and our minds filled with white noise. We were all in a kind of limbo world, a neutral zone.

In the way I had come full stop after giving birth to Cleo, I was unwittingly following the prescription of ancient cultures whose people were brought up knowing just what to do at such a time of transition: nothing. Big life changes were once acknowledged, honored, and ritualized with rites of passage. People at a

crossroads consulted dream figures, cultivated heightened awareness through chanting, meditation, fasting, and psychotropic
plants, and spent a piece of time alone, in the sweat lodge or on
a spirit quest. Those cultures recognized that when people went
through a major life change, they needed space and time in which
to develop a new sense of self. The time alone, empty time, was
a necessary prerequisite to the awareness; it was a period not of
seeking, not even of waiting, but just blankness. People of old
entered that zone consciously; they respected the pause between
an ending and a beginning and treated it as sacred. For me, the
amorphous goo of early parenthood provided a built-in gap
between stages, the opportunity to disconnect and just wait, as
change found form within me. For me, this was not a confident,
deliberate process but a subliminal, instinctive struggle for balance. I felt my way through blindly.

I didn't need psychedelic drugs to put me in a dream state.
Dozens of peptide hormones and sex steroids were doing an
about-face in my bloodstream. The members of the biochemical
power team that had supported embryonic Cleo through her
gestation—estrogen, progesterone, lactogen, and human chorionic
gonadotropin—had all reached peak levels late in my pregnancy,
and were now drained out of my system in one fell swoop. Sleep
deprivation lowered my blood sugar. Night nursing interfered with
melatonin secretion in my brain, leaving me poor in dream sleep
and even, ironically, vulnerable to neurofibrillary tangles just like
the ones Alzheimer's disease was giving Ma. Cortisol, the stress
hormone, had run high during my pregnancy, probably protecting
little pre-Cleo from the onslaught of my immune system, but now
that she was out, it dropped to lower than normal levels. With
luck this would keep my blood pressure low, my immune system
strong, and my response to stress at a minimum. In this way my
body protected me from the normal effects of sleep deprivation,

which under different circumstances could have caused the kind of elevated evening levels of cortisol that are typical of aging and depression.

Meanwhile, hormones also pulsed into my system as a response to Cleo's insistent tugging at my breast, and upkeep of the milk supply exacted a huge energetic price. In my hypothalamus, which controlled hormone secretion in my brain, cells began to rearrange themselves to facilitate the important new role of my old friend oxytocin, triggering milk letdown whenever Cleo noisily suckled. Glial cells called astrocytes (so named because their many little arms reach out in every direction, so they resemble stars), which had surrounded the oxytocin-secreting neurons, began to peel away, allowing other neurons to connect to them and stimulate them. Once again I was under neurological reconstruction.

After giving birth to Zoë, like most postpartum women, I'd had little room for a pause, let alone the vision quest that might follow; I had to get back to my research. So I fought hard *not* to slow down. I deliberately defied every bodily and emotional impulse in my system in order to get back to work, which led me to those afternoons in the abandoned lab at Brandeis, writing in my journal with tears streaming down my face while I pumped milk between electrophysiology experiments. This time I was more fortunate: at home with both kids for the summer, I didn't have to switch gears so abruptly. I had the opportunity to try something new: staying still. Of course I did not literally stay still. As any parent of an infant/toddler pair would guess, I was constantly in motion. I walked the stroller, changed the diapers, shopped, pushed the swing, bounced the bouncy chair, folded laundry, cooked, cleaned, and finally planted those seedlings I'd daydreamed about while I was still at work—but through it all I let Zoë and the baby dictate my every move, and I let myself feel what I was feeling. Emotionally I stayed in one place, while my life changed.

The process of transformation in my body, as with all new mothers, continued for months with everything in flux, but like so many others before me, I found sleep deprivation to be particularly mind-altering. No wonder—no one should ever underestimate the biological effects of interrupted sleep. Many tasks are best learned if the training period is followed by a solid night's sleep. Sleep has been shown to help with several different kinds of processing, from insight formation and novel language perception to visual discrimination and motor skills. New memory traces remain fragile between training and testing unless the subject gets a nice long snooze, whereupon the memories seem to consolidate and then can be retained for weeks or years. Daytime naps help, too, but total continuous nighttime sleep seems to be the most important factor in learning these tasks. Some neuroscientists think that temporary memories form during the day and the hippocampus holds them transiently until we sleep, at which point they are communicated to the cortex for long-term storage. I wondered if this meant that I was being held by my disrupted sleep schedule in a limbo-learning state, where fewer new permanent traces could be laid down—a sort of biologically enforced flexibility.

That was certainly how it felt. In a way, my mind was starting to resemble Ma's; both of us were being forced into a kind of involuntary meditative state, increasingly restricted to the present tense. Her brain, too, was less and less able to process short term memories into long, so what happened a moment ago was more and more readily lost to her; she had memories of long ago, and she had Now. Meanwhile, I was constantly held in "ready" position, with all my energy on tap, continually redirected by Zoë and the baby, and this kept me living in the moment as well. The term *open-minded* means something else, but at that point my mind was definitely wide-open, receptive, as unburdened as a child's. I was relearning how to focus on simple, immediate things that mattered; the constant attention to the needs of my two girls induced a kind of surrender in me.

I didn't know it at the time, but this was the perfect place for me to be as not only my body but my life underwent major restructuring. After three weeks of enforced postpartum blankness and limbo, the next natural step for me would be a period of conscious seeking. I had no spirit guide. My mother would have been the obvious guide to lead me through this particular transition, but Ma was becoming as needy as my two children. Yet in the next several weeks I would consciously come to terms with the fact that I was in transition, not only biologically but psychically, not only from being mother of one to being mother of two but also from being Ma's dependent to being her caregiver.

Two weeks in, when my sutures had barely begun to heal, Ma's phone calls became more frequent and more urgent; she was lonely. I encouraged her to branch out, to join the senior center and in particular the Alzheimer's group there. I sensed that I needed other people involved in Ma's care. The next step would be looking for in-home care or an assisted-living situation, but she wasn't really ready for all that yet—or so I told myself.

At first she was doubtful. "Those senior citizens?" she said. "They're a bunch of strange old people, aren't they? What would I want to do with them?" I knew what she meant; secretly I found the idea depressing myself. But at her core Ma was a social being, an active person, and she liked doing things out of doors. And I needed her to have a life apart from mine. When I pointed out that there was a walking group, an exercise group, and a bocce ball group at the center in addition to the Alzheimer's group, she finally relented and walked down one afternoon to check it out.

Two weeks later I got a call from the leader of the senior center's Alzheimer's group.

"Your mom is such a very sweet lady," she said.

"Yeah," I said. "I think she's pretty sweet." I was waiting to hear what would come next: how she was worried about Ma and thought she really needed more help at home.

"And she's definitely the highest-functioning one in the group. Do you know, she's signed up for walking, exercise, the book club, *and* bocce ball? And she's very conversational. She's really just fine still, isn't she? I'm not sure I would even have realized she had Alzheimer's if she hadn't told me herself."

Hearing this pulled me in two directions. *Well, no, actually,* I wanted to tell her, *she's not all that high functioning anymore, and she's definitely not fine. You just don't know her, you didn't know her before.* At the same time another voice said, *Ha, you see, she's good. This thing is going very slowly; she doesn't really need so much from me yet.*

I needed to believe that for the moment. I wasn't ready to approach Pat with the news that I needed to spend *more* time with Ma. As the summer approached, Zoë was looking forward to kindergarten, Cleo was beginning to vocalize, and Ma at least wasn't in crisis, but in spite of his new job location in Berkeley, Pat was overwhelmed. Following a week or two of sweet new-baby snuggling, he was instantly reengulfed by work. Although there was no doubt he loved us, that was how it felt now: we three girls were an "us," and he was separate. He still came home late most nights, by which time Zoë was usually ragged, and after a full day and evening alone with her and the baby, so was I. Often I'd fall into bed the moment both girls were asleep and never really see him.

With our schedules at such odds, Pat and I spent almost no peaceful downtime together during the week. I missed him, but at least I got the affection of my girls—between Zoë's tight hugs and sloppy kisses and little Cleo at my breast, I had constant contact and unconditional love. His only real contact with any of us happened on the weekends Ma didn't come. I was still keeping her at bay every other weekend, and I felt horrible leaving her alone, but I did it to protect my own little family; weekends were our healing time, our relief. I saw Pat reach out to Zoë, engage her in conversation, try to teach her about electricity, history, numbers; we strolled around campus or met friends in Chinatown for dim sum. Pat and I held hands; we discussed the news, browsed bookstores, and wished there could be more time in a day.

As soon as Ma came back into our lives, which was going to happen soon, the weekends would be shot. Not to mention what would happen in the fall. I lived in fear of the fall semester. How was I going to take care of Ma *and* the family—*and* return to work?

Not only would Ma have a bigger impact on us when I returned to Cal, but we hadn't figured out yet who would do everything I had been doing when I was no longer in the house all day. Since the beginning of our relationship, Pat had done less than I had around the house. I had managed, barely, when it had been just Zoë who needed my care, but I knew I couldn't continue to do all the housework with two kids and a job—not to mention Ma. I had no interest in playing Super Mom, and I shouldn't have had to. I began to take stock. I began to notice how much more I did and to point it out to him.

"Okay," I said one Tuesday morning, "today between nursings and cleaning the house, I will do the shopping, take out the recycling, and run a couple of laundries. All stuff the two of us will be doing together in a couple of months."

He looked me up and down and shook his head silently.

"What?"

But I knew what that look meant: that this was not a good

time to have a conversation about household chores. There was never a good time; he always said the same thing, night or day, week or weekend: "I can't talk about this right now." I had heard those seven words again and again.

"We have to think about this stuff, Pat. We have to. I don't think you realize what we're in for, here."

I was going to fight for this; it was one of my least favorite things about us. Our unequal earning power should not translate into an inequality at home. His job *was* important; we needed it to live, but that didn't make me the maid.

"I can't talk about this right now," he said. "I've got to get to work." He snapped his laptop shut and moved to the door, but then turned back. We had a policy of never leaving each other without a kiss. He stepped quickly over to me and pecked at my cheek, the driest bare minimum of a kiss.

"Don't force yourself," I said, and the door slammed shut.

In the ringing silence that followed, I was left with the task of shutting doors in my mind. I shut the door to He Doesn't Love Me, and to I'm Lonely. Then I shut out the demands I had made: a baby by the time I was thirty-five, the earliest time he could possibly imagine having children; a second one before he was really ready; my sick and needy mom in our house every Friday, disrupting our lives. I refused to face the thought that Pat was in it because I had dragged him there. It was unthinkable to me that my family and our babies—the most important things in my life—might push Pat over the edge, so I continuously held at bay the fear that maybe he had married me before he was ready.

On the way to preschool an hour later, Zoë asked me why Daddy was mad, and I told her that since I was going back to work soon and Daddy didn't have the big commute anymore, I wanted him to do more around the house. She said she wanted him to do more things, too. Half joking, I said, "And I'm going to win the argument!" Zoë nodded slowly and seriously, considering this, and then told me, "When you win the argument, Mama, I'm going to give you a prize."

In the end Pat agreed to prove that it would work. We drew up a chart: three nights a week he would do the dishes and I would make dinner; two nights a week it was the other way around; I would do the grocery shopping one week, him the next. But by the morning after his second dish night, the dishes were piled in the sink; he forgot to shop; and on his nights to cook, he bought spicy takeout food, and I ended up having to cook separate meals for Zoë. We had forgotten to even address the issue of pickups and drop-offs, which I did as usual, but I had a fairly solid idea of how it would have gone if I hadn't. I had won the argument, but I didn't get a prize.

Finally the facts clicked into place. Clearly Pat was unable, or unwilling, to take on his share of the work at home. Also he was right; I didn't bring in that much money. I sat down one night and performed a simple calculation, weighing my salary against the cost of full-time day care for both kids while I worked. The result was shocking: after the day care costs and including the value of my benefits, I would clear exactly $1,000 in *a year*.

Around this time Daniel informed me that Ma was "turning into a recluse." He said that although she was still going down to the senior center for exercise class, she had stopped going to her book club, and he wasn't sure about bocce ball. He'd had to practically force her to go get a haircut, and he wasn't sure how well she was eating, so he'd started taking her shopping whenever he went. Obviously Daniel had been doing my job, and he was politely, gently informing me that this was too much for him alone. I knew I couldn't keep avoiding her. My body had healed; my energy was returning; it was time to step back in and take charge. This gave fuel to the idea that had been slowly forming during my maternity leave: maybe I should take some time off work.

Ever since Zoë's birth, whenever I had complained about a job or about missing Zoë, Pat had maintained that if I really wanted to stay home with her, he was sure we could make it work. I had held on to that idea as one holds on to any escape fantasy, not intending to act on it but just as a pleasant thought to help me through a particularly hard day. "If I really get sick of this," I had told myself, "I will just quit—tomorrow." Then another day would pass. Neither of us guessed I'd ever actually do it.

It had begun as a gut feeling, I told him a week later: I yearned to stay with this baby. I hated the idea of shy little Zoë staying late after kindergarten at her new school in the fall. The reality at home, he had to admit, was that I did the lion's share of the housework and child care, so I'd essentially be holding down two full-time jobs, one at home and one at work, not to mention the fact that Ma was needing me more and more. And since day care for two would cost us my entire paycheck, I no longer had any financial incentive to work at Cal. So why was I heading back there in the fall? Why not do one job for the same price and give our children the benefit of a parent at home?

Pat gave me a long, tired look. He agreed that it made sense financially and in terms of my mental health, but he was also visibly envious. *He* didn't have time to take a spiritual journey and reassess his priorities. He couldn't just decide to quit, and we both knew it. For a moment I saw us as from above: me, struggling to be a good mother, daughter, and wife and bound to fail somewhere; and him, overwhelmed by Lockhart women young and old, with his wonderful California adventure now mired in responsibility, duty, and a lot of hard work. But couldn't he see that this wasn't my fault? I hadn't willed Ma to be sick, I hadn't chosen this new job for him, and it wasn't my fault I didn't make as much money as he did. Besides, what I was proposing *would* be work. A single-dad friend at Zoë's preschool had told me he'd give anything to have a partner who was willing to stay at home with his girls. Why couldn't Pat appreciate what I was offering him?

"All right," he said with a sigh. "I guess you should do it."

The instant he said it, I panicked. Stay-at-home mom. What was I thinking? I didn't say anything, just sat there looking at him, still full of tension. I sat perfectly still, waiting, because I had the feeling he wasn't finished speaking. My shoulders and chest tightened; I couldn't get enough air. God, what would my supervisor at Cal say? What would people at Brandeis think? *She dropped out, she wimped out, she's just a mom now.* A world of emptiness opened up around me; more blank space. Was I copping out? Would my brain turn to mush? Had it already?

"But you'll revamp your CV?" Pat asked. "You'll look for a another job next year?"

"Oh my God, yes. And I still want to lecture." Cal had asked me if I'd fill in for one of the professors going on sabbatical in the fall. "It's not like I want to drop out completely."

CHAPTER 16

Fourth of July

Every morning the first thing I did was to lean out of bed and peek down at Cleo. By the end of June, at close to three months old, she already almost filled her bassinet. She was usually quiet when I awoke, but as soon as she saw me, her whole body wriggled. She beamed her delighted "I love you, Milk Lady" smile and kicked both legs up the air, bringing them down together in great bassinet-shaking thumps. She continued to be a voracious eater. She was also beginning to grasp objects and bring them to her mouth and would spend several minutes at a time in deep concentration, trying to direct the soft end of the pacifier to her lips. And she was busy learning to imitate: when we smiled, Cleo smiled back; when we stuck out our tongues, Cleo stuck out hers. She even seemed to imitate the sounds we made, already on the path to words and language.

For her part, Zoë was suddenly an assertive, self-sufficient four-year-old, putting on her own clothes and shoes, brushing her teeth, and trying to fasten her own seat belt. We had a special box where we deposited marbles, small toys, coins, and other "choky objects" we found around the house, and Zoë was very diligent about eliminating these from Cleo's floor space. (She couldn't crawl yet, but we wanted to be ready when she did.) Zoë held Cleo, fed her, and made her laugh like no one else could. And along with all this responsibility, she was becoming much more aware of other people's feelings, most of all Cleo's. As they lay together on the floor, Cleo patterning her first smiles and early sounds after Zoë's, every flicker of change in Cleo's tone of voice registered in Zoë's attentive face. I realized that this play wasn't just about getting a reaction out of the baby; Zoë was genuinely involved in Cleo's world, sad when Cleo was distraught, and deeply content when she was happy. They lovingly studied each other, provoking, emulating, and learning to anticipate each other's actions and intentions. It was like watching someone unself-consciously playing at the mirror. Each at her own level, our two girls were continually exploring the boundary between self and other.

Around that same time, the scientific world was taking notice of a new class of neurons in the monkey's premotor cortex, neurons that would come to be viewed as key players in all these processes—imitation, language acquisition, and understanding the intentions of others. These were the "mirror neurons." The mirror neuron's claim to fame was that it became active when a monkey performed a particular action, *and also* when the monkey *observed another* performing that same action; not only as Jane grasped the food, but as Jane watched *Mary* grasp it. This raised the tantalizing possibility that these cells, which have since been demonstrated to exist in human brains as well, might be involved in one of the most elusive and attractive primate traits: empathy. The ability to take someone else's perspective, to identify with another's feelings, the way Zoë was beginning to do with Cleo, has clearly been an adaptive trait in primates, who have depended extensively on social interaction. Now mirror neurons offered a possible explanation for how tracking someone's mental state and matching it to resonant states of our own might take place at the physiological level. And I couldn't help thinking that this was exactly what Cleo and Zoë were up to, what we were all up to, during the early days of Cleo's life: syncing up our brain activity so we might have a better understanding of one another as we moved through life together.

For Cleo's first Fourth of July, Pat and I planned to drive out to Ma's house in Martinez, maybe take a swim in her mucky old pool, and walk down to the marina for the traditional small-town fireworks. When I called to check in with Ma, she said she needed to ask me something.

"Okay," I said. "What's up?"

"Wait—I have to find my notes."

I waited while she searched. These were the moments that required me to actively practice empathy. I reminded myself to be patient. I reminded myself that it was hard for her, asking for help like this, after so many years of being the one in charge. I could hear her moving things around on the table, the sound of papers being shuffled. Finally she retrieved the note in question.

"Right, okay. What is our cable company?"

"Hmm," I said. "I think it's Comcast. Right?"

Yes, she thought that was it.

"Why do you need to know, Ma?"

"I don't know. Can you think why that would be important?"

For decades Ma had kept meticulous records, filing important documents alphabetically and chronologically so she could easily retrieve them when needed. Now, teetering just this side of self-sufficient, she had a new system: she recorded each important fact on a small piece of white notepaper and kept it close at hand.

The papers accumulated in drifts around her house and

wadded up in the many pockets of her cargo pants. They told her when she had last showered, what medicines she had taken, what books she needed to return to the library. One was a reminder from me to get a new coat for god's sake, the one she had was ripping out at the seams. Increasingly, though, even when she wrote a reminder, she couldn't make sense of it later.

When Ma had these moments of confusion, she blamed herself. "Stupid, stupid," she said. "I should keep better track of things."

I regularly told her, "Let me help you, Ma. It's a disease. It's okay to let people help." *She's not crazy,* I thought, *she's just sick.* I tried to comfort and support her; I felt a desperate need for her to feel loved. But it was hard; she was working very hard to compensate for short-term memory loss, but it was a losing battle, and she always blamed herself.

At her first annual checkup at the Alzheimer's clinic, tests had shown some improvement in her working memory, which they attributed to her no longer drinking, but also a significant decline in memory, abstract reasoning, and executive function (problem solving, cognitive flexibility, and sequencing). They said this was consistent with their original diagnosis of Alzheimer's disease. Although it was still possible that she had frontotemporal dementia as well, I was beginning to consider my search for an alternative dementia a depressing waste of time; FTD was just as debilitating and incurable as Alzheimer's. Any understanding I had gained from that research was a flea to the mountain of my powerlessness. Ma was losing ground. She was remembering less, and she was less herself every day, and that was hard no matter what.

At the Alzheimer's clinic, I asked the doctors about other medications Ma might take besides Aricept. I had heard about a new drug called memantine that was awaiting FDA approval, and I had fantasies of flying to Germany to smuggle back miracle memory pills for Ma. Memantine worked in a totally different way from Aricept. Whereas Aricept restored a dwindling neurotransmitter to acceptable levels in the brain, memantine protected brain cells against a chemical that was given off in excess by damaged cells. This chemical, an amino acid called glutamate, also normally acted as a neurotransmitter thought to be involved in learning and memory, but when the damaged cells in the Alzheimer's brain released too much of it, it caused further destruction.

Normally, when glutamate crosses a synapse, it causes calcium ions to rush into the cell. It is the influx of these charged molecules that starts up a new electrical impulse to be passed along to the next neuron. When too much glutamate stays in the synapse for too long, however, the calcium flows freely into the cells for longer than they can withstand. The chronic overexposure to calcium wreaks havoc in the cell; the excess glutamate in essence excites the neurons to death. Memantine was designed to curb this phenomenon, which is known as excitotoxicity. It does so by blocking the sites where glutamate normally docks when it arrives at the surface of the receiving cell. This puts a stop to the flow of calcium and in turn to its toxic effect. The drug doesn't bind too tightly, so normal, healthy glutamate signaling is possible, but the toxic levels of calcium are not.

To me, this implied that memantine might provide more than just a Band-Aid; it might actually prevent some permanent damage to neurons. However, when I brought it up, the doctors just shrugged. "We've heard about it," the neurologist said, "and the studies look interesting, but it's no miracle cure. It does seem to help a little, and it seems to do so somewhat later in the disease than Aricept. I do hope we'll get to try it with some of our current patients, like your mom, when it makes it onto the market here. But the effects I've read about are not particularly dramatic, and like Aricept, it doesn't stop the progression of the disease."

I should have known better. In grad school I had read grant proposals that seemed to promise a cure for cancer, if the government would only fund the project; mediocre data could intimate miracles without ever telling a single outright lie. Why had I expected anything different from a pharmaceutical company? Still, I continued to scan the alternatives.

Most available Alzheimer's medications seemed to be acetylcholinesterase inhibitors like Aricept, but other drugs were under investigation. Much of the research was focused on stopping amyloid plaques from damaging brain cells. Some drugs were directed at reducing the amount of beta-amyloid, and others were designed to bind to it and block its effect. Vaccines had been designed in an attempt to train the immune system against the amyloid, and one in particular had looked very promising: patients' immune systems mounted a response against the amyloid, and human test subjects proceeded to perform significantly better at memory tasks than those who had received a placebo. Some of the articles I read gave me visions of Ma at the clinic, smiling brightly as a benevolent doctor placed a pink sugar cube on her tongue or stamped her arm with a special needle, with flashes of newspaper headlines in the background touting the miracle of the new vaccine: "Millions Saved!" Unfortunately, a small percentage of the patients in the most hopeful study suffered from a swelling of the brain much like that seen in viral encephalitis, and this caused such a scare that the drug never made it to market.

I wondered why nobody seemed to be interested in tau, the protein responsible for neurofibrillary tangles. Just because the tangles of tau seemed to emerge after amyloid plaque formation, I thought, why must we assume amyloid is the only perpetrator here? But ever since the late 1990s, when so-called Alzheimer's mice were found to be protected from mental decline after vaccination against beta-amyloid, the majority of researchers had assumed that beta-amyloid was their culprit. What if the two interacted? I wondered. What if both of them contributed to the disease in humans? Part of me wanted to jump back into the laboratory and attack this problem myself, and I wondered how many neuroscientists in the field had landed there after a relative was diagnosed with Alzheimer's, but for the moment I had more pressing matters to attend to.

When we arrived at Ma's house on the Fourth of July, Pat immediately disappeared outside to check the grounds. Pat was funny about our responsibility for Ma. When she visited with us in Berkeley, he often acted as if she were a burden, withdrawing and leaving me to attend to her and the kids; but in Martinez he was always somehow more available to help out, happy to replace fire alarm batteries and change the air filter for the furnace. Maybe this was what he could contribute, I thought: practical things in the physical world. I felt a little rush of gratitude. He was being the "good man" Ma thought him to be. He *was* good. He also always sneaked in and checked the liquor supply to make sure she wasn't drinking.

It was ninety-five degrees and dry out, and Zoë was cranky and thirsty. I opened the refrigerator to scan for juice, and a putrid smell immediately filled my nostrils. The air was warm, though the motor still seemed to be running.

"Uck, Ma, what's up with the fridge?"

"What do you mean?"

"It's warm inside."

"Oh, you know, I did notice it seemed to be a little warm the past few days, now that you mention it, but I thought that must just be normal."

"Normal? Ma, look at this, it's not cold at all. And it stinks—things are spoiling."

She stood behind me peering in at her skim milk, grapefruit juice, some withered green beans, a hunk of cheese, and some meat—pork chops, maybe. That must be the source of the smell. She didn't say anything.

"Ma, it's a refrigerator. It's supposed to keep things cold."

"Well, I guess you're right about that," she said with a sheepish smile. "Do you think I need to call the repairman?"

"I think you need a new fridge."

I asked myself how long this had gone on. The thing could have died suddenly, but observing her confusion, I wondered, what if it had been malfunctioning for much longer? Would she even have noticed?

More disturbing was her reaction when I went to toss the mayonnaise and pork chops. "Oh dear," she said, "that's such a waste of good food."

"But you wouldn't have eaten it—right?"

"When mayonnaise spoils, that can be dangerous," she said. "But maybe I could have just cooked those pork chops tonight. I mean, I only bought them two days ago."

Fortunately I had brought leftovers from the night before: cold chicken and salad, watermelon and cherries, packed into our big red cooler. We ate in silence, all of us left to our own thoughts.

Ma finally looked up. "This is delicious, honey. You are such a good cook. But you shouldn't have gone to all this trouble."

I met Pat's eye with a grim smile. I knew we were both thinking: *It was a good thing I had, or we'd be eating putrid pork chops.*

It was Wednesday, which meant I'd probably be able to have a new fridge installed that week. I told Ma I'd try to get one in the next day, and she didn't give me too much of a fight about that. I'd recently convinced her to let me take over her finances. I had discovered that she was sending multiple large checks every year to several charity organizations, and when I had asked her what they did, she couldn't answer. "But I know I pledged money to them, because they keep on sending me the bills," she had told me, "and I can't say I will donate money and then not do it, now. That would be dishonest."

I'd had her collect all the mail for me for a month and then counted up the solicitations. More than twenty organizations had been preying on her, many of them sending quarterly donation forms that looked just like utility bills. When I looked back over her checkbook, I found she had been giving away over $7,500 a year. All of them sounded like causes Ma would have supported—peace coalitions, homes for battered women, nature conservancies—but they sent Ma these bills, and she blindly made out her checks every month, signing away the teacher's pension that should have been socked away for the expensive assisted care she would need soon, and possibly for decades to come. Now I wrote out checks for the utility bills, threw out the donation forms, and balanced the checkbook, while she withdrew cash for her personal expenses. She missed picking up the mail, which I'd had redirected to my house, but the simplicity of the arrangement agreed with her; balancing her checkbook had become next to impossible.

After dinner Zoë wiped her hands, got up from the table, and walked over to the old chipped and out-of-tune piano. She plinked and plonked a little until she found her way to a tune—slow and simple but impressively lyrical, almost like a Keith Jarrett song. The air stirred a bit, and the round shade pulls swung gently at the open kitchen windows. Ma reached for our dishes, but Pat intervened.

"No, let me get those. You two talk." There he went again.

This time he seemed to be consciously releasing us into the emotional terrain while he attended to the physical world. I cocked my head and squinted at him, suspicious of his motives because he rarely offered to take the dishes at home, yet again I felt grateful that he was there and making things easier for me.

I asked Ma to come out on the little balcony that ran along the left side of the house. Since it was too narrow for sitting, it served mostly as a place to stand and hang laundry from the pulley line that ran out over the backyard, but if you stood at the very end, there was a view of downtown Martinez. I noticed that the railing was getting a little shaky as I stood next to Ma, gazing out at the familiar skyline.

The house perched on a hill that backed up against the old Shell Oil Refinery, and it had a view of the little downtown nestled below us, the high hills on the other side of town, and a glimpse of blue water, the Carquinez Straits. The Fourth of July parade had passed down Main Street that morning, and now downtown, with its county buildings and antique stores closed for the day, was quiet except for the occasional rapid fire from a stray string of firecrackers. Snatches of music and the sound of an amplified announcer's voice wafted up from the marina, where the festivities continued with a small fair and many family barbecues. A train whistle sighed in the distance, and a breeze carried the sound of the clanging of the crossing gate down by the station.

"Ma. I need to talk to you about some things."

"What is it, honey?" She knew by my tone of voice that I meant business. It was time for a serious talk, time for me to bring up the issues on my checklist that we must face together. I had recently realized that she was overdue for a physical, a Pap smear, a dentist appointment, and a bone density check, plus I wanted her to see the assisted-living homes I had looked up, if she wanted to. I'd been waiting for a quiet moment to bring these things up.

"Well, a couple of things. First, I scheduled you a checkup."

"Oh, all right. You'll have to tell me the date."

"I already wrote it on your calendar. There wasn't anything on that day. Okay? I'll call you the night before to remind you, and I'll come out and drive you to Kaiser."

"Okay, honey. That's fine." Her voice was resigned but soft, as though that hadn't been as bad as she'd thought.

"And also . . ." I began, and I saw her frown.

"You know the folks at the clinic said we should check out assisted-care facilities well ahead of time."

Ma sighed the way she always did when I brought up anything to do with this disease. "You mean those places where they put people who are sick, like I am, I suppose." There was a sarcastic, bitter bite to her voice.

"Yes," I said firmly, though I didn't feel firm. "I want to check a few of them out. Just to see what's out there. And I thought if you want to be in on this decision, you might want to take a look, too." I twisted my T-shirt in my hands, wanting to escape, uncomfortable in my own skin. I was supposed to do this, but it hurt *me* to bring it up. I didn't want to think about it; I didn't want her to have to think about it. But I was supposed to take charge, and if I didn't, *she'd* end up in trouble. Taking on this role, I felt like the apprentice putting on the magician's hat; I didn't know the right words, and I suddenly had a kind of power I wasn't sure how to use.

She gazed past the little town to the horizon, where the sun was sinking low, silhouetting the round hills with their scattered oak trees black against the fading sky. She sighed again.

"I love this place, you know."

"I know."

"I haven't lived anywhere else for a long, long time."

I leaned forward, careful not to put all my weight on the loose railing, and looked up at her face. She seemed very present, surprisingly relaxed, almost her old self. "Not since I was born," I said.

She looked at me. "Is that right?"

"Yup. You and Daddy bought this place from the Deesies the year I was born. You don't remember?"

"I guess I do remember that, now that you say it." She looked off in the direction of the marina. "I still have my walk with Daniel every morning. I feed those ducks two or three times each week, you know." I nodded. Cleo and I had gone with her on the tour around the little duck pond three weeks earlier. Ma didn't just scatter old bread crusts; she drove to the feed store once a month and bought a huge bag of high-grade birdseed that she parceled out for them each week. I'd had a good laugh watching her coddle her favorite mallards and shun the "naughty" geese and gulls and blackbirds who "tried to steal their food right out of their mouths," as she put it. How could I possibly try to convince her to leave?

"Do you really think I'll have to move?"

I forced myself to stay on track, looking her in the eyes as I spoke. "I think things are getting harder for you, Ma. It's hard for you to remember to wash your hair. This refrigerator thing—I don't know if you realize how dangerous it would have been to eat that pork. Things like that—well, maybe not now, but sometime there will probably be a time when you aren't safe to be alone." This was as close as I had gotten to discussing the really ugly aspects of what would eventually happen to her.

I could have mentioned the housekeeping as well. The place was still tidy, but it was filthy-tidy. I had the sense that she had straightened things up long ago and now she simply never moved anything. Everything remained in its proper place while dust and grime gathered on top of it. Once when I had asked her how often she vacuumed, she had brought out a little mechanical carpet sweeper: "I vacuum with this every week." The odd thing was that even as Ma began to resemble a child, forgetting things and exercising bad judgment, she nevertheless maintained certain mature expectations for herself; she cleaned like a child, but she received the information that the house was filthy like an adult. This was an insult to her housekeeping skills, an offense. I pressed my lips together for a moment and let that particular issue go, but still I had to press on. I smoothed out my T-shirt and cleared my throat. I could do this without making her feel criticized. I had to come from a place of caring.

"Ma, people with Alzheimer's forget things—like turning off the stove, like locking the doors—and I don't want to have to worry about you when it gets worse. I want you safe. And signing up wouldn't mean you'd move tomorrow—it would just reserve you a spot for when you need it. There are waiting lists."

"I can be more careful, though."

"Ma—" I began, but there was nothing I could say. For eight years I had lived away from these hills that I loved, and I knew that grief for a place could be as potent in its own way as grief for a person. I understood why she was fighting it.

We both looked away into the distance again for a moment, and then suddenly she turned back, surprising me with the speed and force of her body. The railing shook as she grabbed it, and her eyes were full of tears, full of fear and fury. "I don't want to see any of those places. If I ever have to move from here, I think I'll just run away—to *hell*."

This was Ma, through and through, fiery and defiant and sticking to her guns where it really mattered. This was the Ma I loved more than anyone, standing up for what she loved as she always had. I looked at her as the tears began to trickle down her cheeks. She didn't wipe them away. I felt her core. My whole chest flooded with conflict as logic and duty were pushed down and pure raw love marched over them. I knew she wouldn't run away; I knew she'd eventually do whatever I recommended. But she had named my own fear: that hell was precisely where I'd be sending her. Like her, I couldn't imagine her anywhere else but here.

An hour later we walked down to the big grassy park by the Martinez marina, just as I had done with my parents and sister for the first sixteen years of my life. I hadn't known what to say to Ma about moving out of her home, but I knew that I couldn't continue that conversation. I had just pulled her into my arms and hugged her, and a moment later, it had been time to leave for the fireworks. Now, as we headed out down the hill, I felt heavy and exhausted, though I was trying to keep cheerful for Zoë, for whom fireworks should be a pure good time. I had put off nursing Cleo because I was concerned that she might be frightened by the fireworks. I thought I'd use the same trick I used on airplanes at takeoff and landing, comfort nursing at the time of stress. My breasts strained at the fabric of my bra as I marched along with her in the baby backpack.

Ma had recently developed the annoying habit of walking a few steps behind me. No matter how much I slowed down, she would adjust her pace so as to hover just to my right and one or two steps behind me, almost like a dog heeling. I hated it, but I'd come to realize that this was somehow comforting to her; maybe the following relieved her of the need to keep track of where she was, the fear of losing her way. When I glanced back to check on her, her face was full of worry. She was not looking down at her own feet or mine, like she usually did, but was scanning the street anxiously as though she'd misplaced something. Was she feeling lost?

I imagined how transformed the town must look to her, her usual route to the marina now full of people and cars streaming along streets and sidewalks on their way to the park. I realized I should probably be just as concerned about Ma getting lost in the crowd as I was about Zoë, whom I'd firmly instructed to keep hold of Daddy's hand. I stopped and reached out for Ma, linking her arm in mine and pulling her to my side. This made us walk a little clumsily, bumping hips, but I needed her to feel safe.

When we got to the green, I laid a wide picnic blanket down on the grass for the five of us. It was almost dark, and the crowds kept gathering in more tightly around us. I pulled out a sippy cup of milk for Zoë and watched her snuggle onto Pat's lap. Settling Cleo in my own, I smiled wanly over at the two of them, remembering how I'd always wanted to be near my dad for that first quiet thud of ignition, the hiss as a rocket streamed high into the air, and the explosion of light followed by an exhilarating deep, resounding BOOM!, with the smoke carried away on the wind. I loved the warmth of his body as I leaned back, snuggling in, smelling that intoxicating Daddy-aroma of sweat, beer, and Philip Morris cigarettes.

I reached over and took Pat's hand, and he leaned over to kiss me. Barbecue and firecracker smoke filled the air and mixed in with a hint of cotton candy, fresh-cut grass, and the sweet/sulfur smell of the mudflats carried in off the Carquinez Straits in the soft, cooling night air. It was all exactly the same as it had been decades before, and it was all completely different.

In theory, this was the perfect family activity, low-key and sweet, all of us together on a blanket to see the same fireworks show I'd seen as a kid. In practice, it felt disjointed and sad. The old excitement was there, and it wasn't. I didn't like the smoke, the noise, the drunken men. A large, sweaty man sat down directly in front of us, blocking my view. Taking a swig from his bottle of Jack Daniel's, he lit up a cigarette and casually blew smoke directly back onto me and the baby, and suddenly I felt powerless, vulnerable. The night was so perfect and yet so flawed. Overwhelmed as the first sprays of color lit the sky, I leaned my head down over Cleo's sleeping body and heaved a sigh, letting myself cry while the rest of the crowd oohed and ahhed.

Much later, after walking Ma home and getting her settled, after the drive back to Berkeley, after tucking in Zoë, nursing Cleo, and giving Pat a quick kiss before he sighed and fell asleep, I lay awake, willing myself over and over not to clench my teeth. I thought about the warm meat in the refrigerator. I thought about Ma's notepads full of lists: on her dining room table, reminders to water the yard, return the library books, call my sister in Seattle; next to the bathroom sink, a checklist with the dates of her last bath, shampoo, and dose of vitamins. If only she lived just a little bit closer, I could check on her every day . . . but not with a baby and a toddler and a lectureship at Cal about to start. For the hundredth time I shamed myself for not offering to take her into my home. But where would we put her? All her belongings? What would she do all day? And would Pat and I ever survive that as a couple?

My jaw began to ache, and once again I unclenched my teeth, inhaled through my nose, and breathed out, slowly and gently, through my mouth. I thought I knew the answer to the last question, and I wasn't willing to risk it. *Selfish, selfish,* one voice said. *Realistic,* another replied. I had to find a compromise. I could not in good conscience put Ma in a home now; nor could I leave her completely alone in that house. She really wasn't *that* bad off; she wasn't wandering, and she hadn't left the stove on—yet. I resolved to look at homes by myself and get her on a waiting list regardless. Perhaps in the meantime we could find someone to check on her a few times a week—and possibly do some cleaning. This might tide us over for a few months, maybe even a year. We could certainly use the money we'd save, down the line.

CHAPTER 17

Take Care

The following Monday morning I visited the first home. It was located on a busy street in downtown Berkeley, only a mile from my house. I wanted Ma close by, so I could visit often, and this was the nearest place I could find; I went with high hopes. I located the address I'd copied into my notebook and found parking right in front. *Convenient,* I thought, quickly checking my reflection in a storefront: an unpretentious, trim, thirtysomething woman neatly dressed in casual cottons, carrying a baby. Respectable yet approachable; good. In some not quite conscious way, I was half reasoning that if I made a good impression, the people who ran the place would like Ma better, too, and treat her well. With Cleo tucked into a sling across my front, I had both hands free to take notes. I straightened my glasses, tucked a stray hair behind my ear, and headed in through double glass doors.

I stood in a large room with fluorescent lights and a dirty Formica floor. There was no obvious front desk or office, and no one greeted me at the door, although I had made an appointment for a tour. In the lobby to the right of the entrance, two dozen senior citizens filled three rows of plastic chairs, watching a TV mounted high above them. At first I thought the area was a waiting room; it didn't look comfortable enough to be a permanent living space. One woman slowly knitted, and two or three men caressed the handles of their canes, but for the most part the people sat perfectly still, heads tilted back to see the screen or forward as they napped. Every chair was full, and more people stood in a row behind them.

Despite the glare from the front windows and the flash of cars whisking by on their morning commute, the residents all stared as a perky housewife demonstrated how her cleanser chased the greasy pizza stains from her husband's T-shirts. Every thirty seconds or so a tall man with pursed lips in the back row clapped his hands once, loudly. Otherwise they all remained silent.

When I finally located an attendant, she spoke to me in a gruff voice, as though I had offended her by merely showing up. "Well. It looks like you've already seen the living room," she said, crossing thick arms across her large chest. She was squat and solid, with short brown hair, bored, tired eyes, and bad dandruff. "I guess you'll want to see the quarters as well." She directed me up a long, dark stairway to the dreary hall designated the "Alzheimer's ward." It smelled of urine and disinfectant. I peered into each of five barren rooms. They all had yellowing, cracked paint and cold metal institutional beds. Bathrooms were separate, each one shared among four residents.

As I walked through, I felt my shoulders tense, and without thinking, I reached my arms out to encircle sleeping Cleo, as if to protect her from what I saw. This couldn't be real. I pictured the exposé on *60 Minutes:* "Forgotten Seniors." I continued to take notes, not because I thought this was a possibility for Ma, but in a sort of defense against the dread that was welling up in me; instead of turning around and running away screaming, I took the anthropologist's approach, taking notes with a level, unbiased eye, giving the mundane and the morbid details equal weight, but underneath I was terrified. The attendant told me in a bored monotone that shared rooms cost $2,000 a month, private were $2,500, but there were unnamed additional costs, depending on the level of care required.

Where were their belongings? I wondered. "Residents are allowed to bring their own television, bed, and a small bureau," the attendant told me, "but most of them don't, since we can provide what they need." I thought of the fruit trees and daffodils in Ma's garden, of the path Daniel and Ma took around the marina park ball field at seven a.m., with the birds and the train whistle and the drone of a foghorn in the distance out on the strait. Then I looked out the window, at the roofs of cars speeding by one story below us. What could they possibly provide that Ma *needed*?

On the way back downstairs, Cleo woke up and began to struggle in her sling, so I slipped the notebook into my backpack and brought her out. We were on our way from the stairs to the "living room," walking down a dark hallway that also opened onto the kitchen. It smelled as if someone had fried fish recently, and it crossed my mind that possibly the oil had been rancid. Suddenly a woman swung out of the darkness. "That's my baby!" she hissed. I stepped back, startled. She had long, wild gray hair, large hollowed eyes, and pale white skin. She came at us very fast, her long bony arms outstretched, and as I tried to step away from her, I felt something—a couch?—hit the backs of my calves in the darkness. She began to snatch at Cleo "You took my baby!" she screeched. "Give him *back*!"

At that point I did run. I pushed past the woman and the attendant. I didn't say good-bye to the attendant, and I didn't stop until I reached my car, even though Cleo, startled, had begun to cry. When I got in, I just joined her, sobbing as she wailed, holding her too tight in the hot, stuffy car.

Tuesday I persevered, though I was emotionally exhausted. I set out prepared for the worst but discovered a very different place—a high-rise in a popular, bustling Oakland neighborhood, right around the corner from a theater and one of my favorite cafés. It was like a play version of independent city living: the rooms looked and felt like apartments except that they lacked kitchens; all the meals were served in a cafeteria below. Residents seemed autonomous, but attendants came and went, checking to be sure they were safe and had their medication, and security at the front door kept track of all comings and goings. *Pat and I should sign up for this one day,* I thought. With cafés and bookstores a short stroll around the corner, it seemed the perfect place for an active elderly person with city habits to grow old. However, there was no Alzheimer's ward, and no garden. This was not a place for my gentle, befuddled small-town mom.

Wednesday, venturing farther into Oakland, I found an even nicer senior living situation. Well-monitored, mostly female advanced-care elders, dressed and groomed but vacant and staring, slumped on wheelchairs in the cheery terraced gardens. Highly skilled nursing staff trained specifically to attend to the needs of Alzheimer's patients hovered in a ratio of approximately one to every two residents. This was the country club of Alzheimer's facilities. The cost after "additional expenses," which they revealed to me only after my hour-long tour, approached $200 a day.

I began thinking much more seriously about money. I considered the possibility of placing Ma in a home where Alzheimer's expertise and skilled nursing were not available, then transferring her to one where they were, since the Alzheimer's experts were expensive, and if we could save up money that way, she might be able to stay somewhere nicer later on. On the other hand, wouldn't it be kinder to get her into a home with graded care now, so by the time the really hard part came, she would be established in a comfortable, familiar environment and not have to undergo the trauma of multiple moves? Then again, maybe none of this would matter to her by the time she reached the advanced stages of the disease; maybe by the time she needed skilled nursing, she would be so unaware of her surroundings, we could stash her anywhere. That thought hurt the most.

A month later, I drove out to Martinez to take Ma to her physical exam at Kaiser Hospital. She was wearing the same dirty, rumpled clothes she had worn the past four times I'd seen her. She wore her greasy blue hat with the earflaps down. Her glasses were speckled with spots of food and flaked skin, and she peered out squintingly, scrunching up her nose as she rushed headlong into the hospital room. She seemed to expect the doctor to arrive at any moment, and she hurried to undress even after I reminded her that this was Kaiser and they always kept her waiting a minimum of twenty minutes. She grunted, knees popping, as she bent to remove each shoe and each loose gray sock. I spotted an odd curling yellow knob of fungus on one of her toenails.

"Contagious," the nurse-practitioner would tell me, "and practically impossible to get rid of. They won't touch her at the nail parlor. Be sure you have her wear socks around the house."

Her fraying cargo pants fell to the floor in a crumpled pile, and she stepped out, then gently folded them. I glanced uneasily about, my eyes avoiding her body. I was acutely aware of her white underpants, the same kind of underpants she had worn when I was ten: simple, worn white cotton briefs.

I tried to remain casual. I had anticipated and avoided this moment, and I dreaded the sight of her old and sagging breasts, her flaccid, spotted flesh. For many months now I had been watching the steady degradation of her personality: the Loss Of. The loss of passion, the loss of opinions, the loss of clarity—the loss of self. I was afraid of seeing my old mom so vulnerable, so weak: naked. At the same time I felt a morbid fascination, anticipating how the sight of her physical being would unearth me; it would mirror my experience of her slow but monumental shift from laughing, competent, assertive woman to passive, ornery, humorless lump. Her body would give me confirmation of her mind's deterioration, I thought, when I witnessed that fragile old skeleton drooping with the flesh of the infirm.

I couldn't resist. I looked. And I looked again, surprised. Her legs were full and alive. These legs had fleshy, muscled thighs and calves, shaped familiarly like my own. And something shifted inside me. She pulled at her bra: an unself-conscious act. Her fingers calmly followed a course set by many decades of habit. This, too, surprised me. Had I expected her to fumble, to blush at the intimacy of revealing herself to me? She seemed to give her nakedness before me about as much thought as Cleo did when I changed her diaper. And there out of her bra emerged two full round breasts. They were several sizes larger than mine, just as they always had been: big, earthy breasts. Why had I expected skinny, wrinkled tubular sacks hanging down to her belly button?

There she stood, a still-attractive, shapely woman. She stooped only slightly. Without her clothes, and seen in this moment of unconscious activity, Ma was fully physical, generously feminine. Her gray hair fell sweetly to her shoulders as she slipped into a blue gown. She climbed onto the doctor's table, tissue paper crackling under her, and leaned back with a sigh. In that moment, I felt drawn to her with warm, complete affection. My mother had grace; she had a natural, innocent composure. Maybe her blurry indifference added to this effect. But her animal body had an ease that her worried mind no longer possessed. Her incessant fretting, her fading memory and muddied thoughts, could not override the simple, beautiful fact of her womanhood.

* * *

Eventually I visited one last home, which was back in Berkeley. It was run by a nationwide assisted-care-facility-management corporation, but it had a very personal feel. The people who ran the place did not bark; the kitchen did not smell like rancid grease, nor the hallways like urine. The residents were neither neglected nor too dressed up; they dressed casually, like Ma. The rooms, although small, were filled with personal items—furniture, pictures, books, and artwork—and many of the residents sat chatting with family members who had come to visit. Dining was restaurant style, and family members could join residents for dinner once a week. In addition to a TV, the shared area had potted plants, comfortable carpets, scattered couches and easy chairs, and a table full of board games; almost no one was watching the television.

My guide showed me a communal photo gallery where each resident could hang portraits of her family and friends, "to remind them who the people in their lives have been to them." She spoke warmly to the residents, joking and touching them frequently on the shoulder or arm. Another attendant walked up and ogled Cleo in her stroller: "We love babies here." She turned to a resident. "Everyone loves to see a baby, don't we, Mac?" Mac nodded, grinning a crooked, empty-eyed grin, and stepped forward to eye Cleo eagerly. "But we only touch them when invited," she reminded him firmly. "You say your mom was diagnosed about a year ago? We have several early-stage Alzheimer's residents at the moment, all very sweet people."

A prickling warmth filled my chest as I walked the halls; I was no longer listening to my guide's words, I was only feeling the place, feeling the warmth. With a huge rush of relief and gratitude, I realized I could picture Ma here in her care—not now, but maybe someday.

That night I called Alice. I told her that I'd looked at some assisted-living places, and one might even be okay, if we could afford it. They charged $4,500 per month for a private room or $3,400 per month shared, with prices as always increasing with the level of care. I was desperate to believe we could make this work. Before my little tour, I had underestimated how much good care would cost. On that first dreadful visit, to the home with the living room to die in, I had unwittingly begun at the bottom of the heap; naïvely, I had not believed that $2,000 per month could be the extreme low end of the scale.

If we were paying upward of $3,500 per month, Ma's $1,980-per-month teacher pension would not even come close to covering her expenses, so we'd have to immediately tap in to her savings. Thank goodness she'd socked her money away so diligently in the decades since Daddy had taken his disability retirement, and thank goodness I had taken over her finances before she gave it all away. If I invested the savings carefully, they might just stretch to cover the next two decades of care. I told Alice how some senior homes required a financial qualification or a major deposit; some encouraged residents to hand their savings over to the home to invest. Alice started talking about the stock market and mutual funds. I realized this would be my crash course in the kind of money matters I had fearfully avoided all my adult life.

I spent a moment acknowledging to myself that Ma would leave no inheritance. A childish part of me reared up, pouting; no more presents, no more cushion. Ma had always had my back. This wasn't about financial need, though; I knew Alice and I would both be fine on that count. It had more to do with the feeling of receiving. Not quite consciously, I had to acknowledge that I would never again be Ma's dependent; she wouldn't be Mommy anymore. Which meant I had grown up.

"Oh God, Alice, these are such hard decisions. How do you

take over the finances of the person who gave you your allowance? How do you choose the right place when none of them is home?"

"You know, I recently noticed a rest home not far from us," Alice said. "I wonder if they have an Alzheimer's division."

"Yeah? Well, look, it might be a good idea to check it out. The cost of living is so much lower in Seattle—maybe we'd get more care for our money. The important thing is that she end up near one of us. It doesn't really matter which, as long as one of us can be close by."

For one beat, I heard utter silence on the other end of the line. "No," she said.

"What?"

"No, I couldn't imagine having to do that."

"What? What do you mean?"

"Sybil, I'm so overextended. There's no way I could afford even one extra hour a month to go visit somebody in a home."

This was news to me. In the past few months, Alice had asked repeatedly what she could do; she had vowed to help in whatever way she could from Seattle. But Alice was the breadwinner in her family; her husband, Josh, was the one in my position, minus the lectureship; he stayed home with the kids in Seattle while Alice commuted to work at Microsoft in Bellevue. Still, I thought, when Pat and I had both been working full-time, Ma had stayed with us almost every Friday night. I felt my temples go hot. Of course it was hard to contemplate, but how could Alice just say no so quickly?

My voice went low and slow. "Nobody has *time* for it," I said. "It's just what family does."

She let out an exasperated breath. "Of course it is. But I can't expect Josh to do everything you're doing now—and I couldn't do it myself, Sybil. I'm so sorry, but there's just no way."

I reminded her that Pat and I could easily end up moving

back east, given that his mom was still in Connecticut. "Then Ma wouldn't be close to either of us," I argued.

Alice floored me with her next question: "But wouldn't you just take her with you?"

Talk about presumptuous! Take her with us? I could feel my twelve-year-old-little-sister self pushing its way forward through the layers of balance and wisdom, like a worm, like a pest. I couldn't stop it. Self-righteous, competitive little me demanded an equal share of the ice cream, equal time in the front seat; I should get just as much as she; she should sacrifice as much as I. My voice was getting shrill. "Oh my God, Alice. Take her with us? Look, you guys aren't planning to move anytime soon, are you? Maybe it would actually make more sense to have her close to you to begin with, if you think about it that way."

I didn't really believe this. For one thing, Pat and I had no immediate plan to move. And as inconvenient, depressing, and stressful as it had been taking on Ma's finances, her checkups, her housecleaning, and her weekly visits, I still loved her so much. In fact, I couldn't stand the thought of her up there in that gray, wet city, far from her rolling golden hills, far from me. I just wanted Alice to acknowledge what I was doing for all of us. It was work, but it was the right work for me to do at that moment. I just wanted to be appreciated.

"I'm sorry," Alice said. I could hear the tears in her voice.

The anger had already begun to drain out of me, leaving my body in a postadrenaline slump. "It's not your fault. You've got a family to support."

"I do. But I feel awful. I know how much you're doing."

I let that sink in. Somehow the words you need to hear can hurt almost as much as their absence; they can break you open, and then you feel the need even more intensely. I felt relief spread through my body—and yet there was still no happy resolution.

"Thanks. Thanks for saying that."

The tears were coming. I realized that had been the first time in weeks that anyone had acknowledged I was doing anything at all. Stay-at-home-mom status carried such a stigma. When Daniel and his daughter, my friend Laurel, had heard I was planning to stay home with the kids for a while after lecturing, they had teased me. Maybe they hadn't meant to hurt me, but their words had seemed cruel at the time. "So you're going to join the Garden Club?" Laurel had breezily asked me as we watched our girls playing together at a preschool party. "What's that?" I had asked, utterly clueless. She said, "You know—stay at home?" Daniel's words had echoed hers a week later in Martinez: "I understand you're going to be living a life of leisure again soon." All I was doing for Ma notwithstanding, since when was staying home with a baby not work?

But Alice got it. She couldn't do what I was doing, but she understood what it meant that I was there with Ma. She understood.

We sat in silence for a few seconds, and I heard her blow her nose. "This is *hard*," she said. "Don't you think it would be so much easier if you and Pat would just move up here to Seattle and we could all take care of her and the kids together?" This was a familiar refrain, and it came from a good place. Alice wanted our families to be closer, she ached for our kids to grow up together, and I guessed she imagined that if we were there, she could find a way to help more with Ma. Yet in this context it felt like one more ridiculous demand; I could only respond with a long, exasperated, tired sigh. I'd spent too much time on the phone. I needed to focus on what was in front of me, not some utopian fantasy. On my list of things to do, the next pressing item was sneakers. Zoë was growing out of all her shoes.

Alice was right that it would be nice to have her and Josh nearby. I imagined discussing philosophy and literature with

Josh at the park while our kids frolicked happily and Alice and Pat pursued their high-powered techie careers. I wasn't sorry I'd chosen to stay home for a year, but I hadn't quite figured out how this at-home-parent thing worked socially. I had been surprised to find that my change in career status made an undeniable difference to other people. On the rare occasions when we attended a social event with Pat's co-workers, people would ask me what I did. In the old days, when I had replied that I taught and studied neuroscience, people would ask me what I thought of the latest Oliver Sacks book or ask if I'd read the *Wall Street Journal*'s review of a hot new biotech company doing neuro research. Now people smiled uncertainly and said, "Oh, sweet. But are you still *working* as well?"

I felt like a nonperson, unappreciated, and relegated to the corner to nurse. Daniel and Laurel's catty remarks still rankled me. Life of leisure, indeed. And the Garden Club—what a joke. Besides being misleading (I was working, harder than ever), the club image itself was inappropriate; I wasn't in any club at all. Socially I was utterly isolated. Aside from a brief stint lecturing on neurobiology, I was severed from the academic world. I didn't miss my old job, but this new work of mothering and daughtering left me lonely. All my friends worked outside the home, and I quickly realized that the bulk of my social contact during the week had been at work. I had a lot to say and no one to talk to.

As an outlet, in brief moments, during Cleo's naps or while she was nursing, I began to scribble manically into my journal. Since the age of thirteen I had kept a journal, and on and off I had dreamed of "learning to write," but this was different. I found myself chronicling my experience with Ma and the kids, pondering the many questions I had been afraid to ask myself: How would we manage? Where would Ma be in ten years? Was I going to get Alzheimer's, too?

The words spilled onto the pages of my journal the night of that first assisted-care home visit. That entry blossomed into an essay about Ma and Alzheimer's disease. When it got too unwieldy for the journal, I switched to one of Pat's discarded laptop computers, and as soon as I had it in print, I knew I wanted it to be read. I felt for the first time that I wanted to write *to* someone; I just didn't know who.

That was when I met Amy, a women's studies professor on maternity leave. I watched her with envy as she tapped away at her laptop on the lawn in the park while the babysitter pushed her toddler on the swing. Every few minutes the little girl barreled back to nurse, or to kiss her mama's cheek, then raced away again to play. I turned to the babysitter and said, "I want what that woman has!"

"You should talk to her," she replied. "Maybe you can have it."

Amy was organizing a group of mother writers to read one another's work online and then meet for a critique session in a playroom full of their toddlers. Shyly, I revealed my essay to these women, who were all like me: moms with active minds, at home with young kids and wanting to express something about the experience.

Within weeks I was hooked. Communicating this way turned out to be the mental equivalent of restoring my electrolyte balance to avoid heatstroke. Writing was not the dark and isolated practice I had imagined it to be. There seemed to be a tangible filament connecting me to every reader and to the authors I read, all of us forming a vibrant web of resonating voices. I felt like a long-idle power station that had just been reattached to the grid.

What adaptive mechanism had kicked in to give me this new connectedness just when I had begun to detach, from Patrick and from the rest of the world? What was unique about the minds of writers, of compulsive journal writers, or of regular folks who could find peace scratching out a weekly letter to a friend?

A soft buzz of recognition warmed me as I read a newspaper article about Sue Rubin, an autistic woman who was believed to be severely retarded and beyond hope until she was thirteen, when some kind soul proffered a keyboard. As she wrote, she awoke; the words gave her a path to the world outside her own mind; they allowed her to reach other people. Maybe my brain had some similarity to Sue's, an overabundance of certain synapses or cells, or a lack of connectivity between them, such that writing had become my path to community. Or maybe I was more like author Alice Flaherty, who, following a personal trauma, had an episode of hypergraphia, a driving compulsion to write that caused her to fill notebooks, walls, even floors with her words.

The causes of autism and hypergraphia are poorly understood, though structures in the temporal lobe of the brain have been implicated in both. I wondered whether the sixty blank books I had filled with journal entries since my thirteenth birthday qualified me as marginally hypergraphic, and whether I'd still be locked in a room inside my own head if I hadn't discovered that writing escape early in life. What I knew for certain was that Sue Rubin, Alice Flaherty, and I formed a continuum. We had in common that fundamental human quality, the desire to join others of our kind, and writing had given us that; writing bestowed fellowship.

I had never gone to the caregivers' support group that Nurse Carol had recommended months earlier, but now I had found a group of like-minded, supportive people. I attached myself to these anchors with a tenacity born of desperation, which I believe is the best glue. Caregivers have to take care of ourselves. We cope however we cope. It almost didn't matter whom I asked for help; the key was in the asking.

CHAPTER 18

Mama in the Middle

Many months passed in this way, with Ma's intellect and personality very slowly deteriorating while the rest of us steadily adjusted, living our lives as normally as we could. Pat continued to work in San Francisco. I continued to support the kids at home and kept on writing. My writing group founded an online magazine called *Literary Mama* to showcase our work, and I began a column there called "Mama in the Middle," in which I chronicled the experience of caring for young children and my aging mother.

By the time Zoë turned six that April, a friend calling me on a Friday night would most likely hear her piping voice at the other end of the phone line first. Without saying hello, she would yell over the Glenn Miller blasting in the dining room, "WHO IS THIS?" In the background the blare of the music would be usurped by the wail of two-year-old Cleo, who had just been robbed of her copy of the *Bay Guardian* newspaper and was going to raise holy hell until her sister gave it back.

A moment later Zoë might pass the phone to me in the living

room, where I'd be with my mom, mid-battle. "No, Ma," I'd be saying, "the last time you were here, you really did wet the bed, and the time before that, too. You have to wear the Depends when you sleep here." The receiver would clunk to the floor and roll around underfoot for a moment before I finally answered, the exasperation in my voice melting into resigned fatigue. "Hello?"

While Ma's social skills had deteriorated, Cleo had begun to negotiate complex friendships in preschool with impressive grace and generosity; her teachers told us she was a "connector," a child who brought others together. Meanwhile, in the fall of 2002, Zoë had begun kindergarten and commenced to read every book within her reach. This happened just as Ma's reading level began to slip, from sophisticated literature to quality young adult literature and finally to dime-a-dozen formula romance novels. And while Zoë began ballet and karate classes and developed a distinctly improved sense of direction, Ma had started to falter on her feet from time to time and to lose her way far more often. During this period the intersection of the girls' development and Ma's deterioration struck me as almost an equal exchange, as if there had been an actual transfer of her former abilities directly to them; as if, peeking inside their brains, we would see their hippocampi and prefrontal cortices growing plump and healthy while hers went limp, lesioned, and depleted.

One grave consequence of these changes for Ma concerned her driver's license. The Alzheimer's clinic had warned us that it was legally obligated to report her diagnosis to the Department of Motor Vehicles, but we hadn't thought much about that until she got into a little fender bender. She ran a stop sign, gently bumped another car in the intersection, and had no explanation. "I don't know what I was thinking," she said. "I just didn't stop." She had admitted it was her fault, and the couple in the other car had been nice enough about it, but the following month the DMV sent her a notice that she was to report to an office in Oakland for a special driving test. Annoyed at this inconvenience, Ma told me, "If you'd never taken me to that Alzheimer's place, I wouldn't have to do this, you know. I'm a very good driver."

It was true that Ma had always been an impeccable driver. As I considered this, the image that came to mind was the back of her head: from the backseat of her car, it was what I always saw. I had memorized the back of her neck, tanned from all the time she spent in her garden; a few stray brown hairs always straggled out of her rubber-banded ponytail. I liked to watch her eyes in the mirror as we talked; for brief moments on the freeway or when we stopped at a red light, they would meet mine as she spoke, but otherwise she never lost sight of the road. When she had to make an abrupt stop, her right hand would shoot out reflexively, as if to keep her front passenger from flying forward into the dashboard. In over fifty years of driving, Ma had never caused an accident, until now. *Who knows?* I told myself. *Maybe she'll do just fine.*

"Ma," I told her, "I know it's a pain, but just go and do it. Lots of seniors have to do this, and they do it for your own safety. I'm sure you'll do okay."

But Ma missed the appointment. She got lost on the way there and showed up over an hour late. The DMV was nice about it; they agreed to reschedule the appointment at her convenience, and for the second appointment, she studied the maps and left her house several hours ahead of time. This time she made it to the appointment, but she still managed to fail the exam.

"What happened?" I asked.

"Well, I did fine with the driving."

"But?"

"But they had this last thing where you had to drive for a while with the man giving directions, and then find your way back without any help."

"And?"

"And I said no."

"You what?"

"I said no. I said there is no way I am doing that. I can't. I can't do that. But I can *drive* just fine. And that man told me I was going to fail if I didn't try. But I knew I couldn't find my way back. So I failed." She spat this out defiantly. Her voice had a familiar quality to it, reckless and irate. *She needs a drink,* I thought automatically, before realizing that was no longer an option.

"Oh, Ma. That sounds so awful. I'm sorry."

"It was. But I should fight this, right? I mean, there must be some way to ask them for another chance, what do you call that again—"

"Appeal it?"

"Yes. Appeal it. I should appeal it, right?"

I wanted to say yes, of course you should appeal it. Ma was, after all, The Driver. She had been the official driver of our family beginning the day Daddy received disability retirement from the county of Contra Costa for his glaucoma in 1967. For our shopping trips to the Sun Valley Mall and the Geary Road Co-op in Walnut Creek, it had been Ma who buckled us in and took the wheel, while Daddy climbed into the front passenger seat, with his white cane beside him, his brown knapsack at his feet, and a cigarette in hand. "Everybody in?" she would ask. "All buckled up? Let's go!" She was a safe driver, but you could see the joy she took in that first surge down Lafayette Street; the car was her freedom, her power, a part of her private world that we entered by invitation.

Ma's little cars—a green VW bug, then a bright red Squareback, and later a little silver Honda Civic—always smelled like the cinnamon Trident sugarless gum she chewed when she got drowsy on the road; her radio was always tuned to KPFA, the local public lefty station, and the backseat was always covered with dog hair. She drove us to the beach, to the first *Star Wars* movie, to the hobby stores where Daddy bought plastic model airplanes and met his nerdy model airplane pals, smoking and arguing over the fine points of war plane insignia. She commuted to work every day for so many years that by the time she was ready to retire, she had rows of little precancerous growths on the backs of her fingers, from all that sun atop her steering wheel; that year she started wearing white gloves for driving.

I thought about Ma's current driving style: fast and frantic. She kept two hands on the wheel, just as she always had, but her knuckles went white; she seemed to be grasping in desperation, just praying the car would take her where she needed to go. She stuck to the roads she knew well, as she must, or else she would get hopelessly lost. There really had been no explanation for running that stop sign. What if a child had been in the crosswalk? Her car was her freedom, her car was her path to me, but I knew what I would think if she were somebody else's mom: I'd want her off the road.

In the end, I admitted that the DMV probably knew what they were doing. Probably she'd be safer not driving. I drove her to the DMV, where she was to exchange the license for an official California ID. The office was in the town of Walnut Creek, where we had shopped every week when I was little. The road from Martinez was busier now—a bank, a pet store, and a bulk liquor outlet had sprung up in the intervening decades, and the hills were sprinkled with rows of new town houses—but the quality of the hot dry air and the shape of the surrounding hills took me back to those days of riding in the backseat of the VW Squareback while Ma drove. I gripped the wheel and glanced over at her, but she did not look back at me.

The DMV office was air-conditioned, the linoleum tiles dirty.

We stood mute, listening to the conversations around us. Right before we got to the front of the line, Ma turned and looked anxiously at the door, as if she might still run back out and appeal those test results—as if there were any way to get another score on that test. Then she turned and looked directly into the eyes of the man behind the counter. I stood behind her and watched as she surrendered the laminated card that had represented her identity for so many years. Without explaining why she was there, she simply held out the license at arm's length, her eyes full of tears.

Afterward we went straight to the feed store, where she bought the most enormous bag of birdseed I had ever seen. This was typical of Ma's perseverance in all things kind and good. She might lose her freedom, but her mallards would damn well be fed.

The loss of her license was one more hole in the fabric of Ma's life. She began riding the Amtrak train from Martinez to Berkeley, but logistically this was a stretch. Fortunately she could walk to the station, and often Daniel was able to see her onto the train. If I was there to meet her at the other end, and the conductor had instructions to look out for her during the trip, the short ride was workable. On Friday afternoons the kids and I picked her up at the Berkeley station, and on Saturday afternoons, after brunch at Saul's deli, I drove her home, often stopping to do her grocery shopping on the way. But we knew that soon the time would come when this, too, would be too much for her.

As her illness progressed, our routines became ever more important to Ma, and the Friday night ritual was down to a science. The girls and I had to be at the station before she arrived, or Ma would worry so severely I thought she might physically collapse—not to mention how I worried that she'd wander off or be mugged at the station. We then had to drive to the Cheese Board for pizza, I had to serve salad with dinner, and that had to have avocado and balsamic vinegar. Her breakfast at Saul's—it had to be Saul's—the next morning was always the same: scrambled eggs and pastrami, a small orange juice. And so on.

With two small mobile children, the house was always full of sound and movement. Saturday morning, shortly after seven a.m., the girls would arrive in our room. I was usually awake, but Pat often rolled over, resisting the noise and chaos to claim forty minutes' more sleep while the kids and I bounced and snuggled and tickled and eventually roused him. By this time my internal alarm usually told me that Ma needed her coffee, and I needed a potty break.

The bathroom, once a peaceful refuge, was now a top theater venue. My girls loved a captive audience. They would burst in, sporting African masks and nothing else, and treat me to a performance of the "naked dance" to the tune of Alvin and the Chipmunks' "Witch Doctor" while I finished up and washed my hands. Then it was off to breakfast, the park, and the grocery store, and then the drive back to Ma's house, unless she stayed another night.

Late in the evening, as I snuggled into bed with my cup of tea and the latest *New Yorker,* I'd hear the clicking of Patrick's keyboard in the study, the thump of a foot against the wall in the girls' room, and if Ma was there and remembered to go, her quiet sigh as she trudged to the bathroom. Then I had about twenty-five minutes of downtime before I sank into exhausted sleep.

I'd like to say I used that quiet time in healthy, restorative ways—meditation, reading, yoga—but in truth I mostly stared out my bedroom window, my eyes following the headlights that winked along a distant hillside, and worried. I worried that the girls would do all the crazy things I had done as a kid but not be as lucky surviving them. I worried that Ma would be swindled by the in-house attendant I had yet to hire, or get lost walking home from the senior center. I worried that Patrick would die and we'd all go insane with grief. I worried, because I cared so much about the people in my life, but no matter how hard I worked, I couldn't protect all of them all the time.

The next morning I'd wake up with *The New Yorker* stuck to my cheek as Cleo crawled in under the comforter and inserted her glacial, wiggling body into the small, warm space between Patrick and me. Then Zoë would claim my other side with a shiver and ask, "Is it a school day?"

I was physically in the middle like this most of every day, but I was also closing in on other personal midpoints. By birth and in spirit, I was perfectly middle-class. When I had the willpower to stay up and write, I did so by the glow of my laptop computer, in the middle of the night. And although I hadn't seen it coming, it seemed I had landed squarely in the middle of the stereotype I had defied for most of my adult years: I was married, at home with two children. My husband made the money while I shuttled my kids and my mom variously to swim lessons, playdates, doctor's appointments, and dementia assessments; changed diapers (wee and grand); tended the garden; did the dishes; and put out the recycling.

Patrick was not happy with the situation. When a year had gone by and I had failed to find a new paying position, I hadn't worried about it—he had found another job, one he liked better and that paid more. We were still able to pay bills on his income, and I could stay with Cleo, work in Zoë's classroom, and tend to the housework and all Ma's business—but Pat saw it differently. We didn't discuss it openly very often, but we both felt the tension. He still didn't like having to be our sole financial provider, and he didn't seem to understand exactly what I was providing.

One Thursday night after a hard day at work, he came home

to find Ma there, a day earlier than usual. "She was *lonely*," I said, "and we were out there helping her in the garden. It just felt right to bring her home tonight." I apologized for not calling to check with him about it, but a few minutes later, I foolishly brought up a topic that had been on my mind for some time: preschool for Cleo. I thought it would be good for her socially to go part-time.

"She's getting too big to bring into Zoë's classroom," I said, "so this way I could keep helping out without the distraction— and I could do more around the house for Ma. And maybe even have some more time to write."

Pat sat silently for a moment. Then he opened his laptop. He didn't look at me as he powered it up, but after a moment, staring off above the screen, he spoke. "You quit your job so you could stay at home with the kids. But now Zoë's in school, and you want me to pay for Cleo to go to preschool—while, what, you go off and do your writing with your friends?"

I looked down at my lap, suddenly confused, my cheeks burning. Was he right? Was I just being lazy? At the time, outside of the uncommon, precious moments when I found time to write, I rarely paused to consider what it all meant. From minute to minute the dozens of small tasks kept me grounded in the present. Zoë had a kindergarten dance performance in twenty minutes, and I still hadn't found the camera. Ma was standing close behind me, fretting about whether to bring her purse. Then, just as we stepped out the door, Cleo felt a poo coming. I was in the thick of it, juggling the urgent and conflicting needs of two short, crazy people and the forgetful gray-haired lady who used to be my mom. If I paused to contemplate my choices, I suffered consequences: Cleo's poopy pants; Zoë's missed recital; Ma's lost purse. I was locked in to the instant; I was a reluctant Zen mother, present for each and every moment.

A swirl of uncertainty spiraled in on me. I had so many

preconceptions about lazy housewives. I'd always been a working person, always supported myself. What was I thinking? Now Pat seemed to think that I had turned selfish, that I was using him to fund my luxurious life at home with the children, and I had to stop and wonder if he was right. Yet through the confusion, there was a burning core of me that didn't believe that. I was more personally anchored than I had been at almost any other point in life. I had discovered that the habit of caring spread from mothering to other parts of my life. I had become much more likely to give the time of day to the panhandling, bedraggled stranger on the street, because my parent-mind transformed him into "somebody's child" or "someone's poor old dad." For once I really knew my neighbors. I'd become more compassionate, more empathetic. I communicated, I participated, I was increasingly, instinctively, centered in my own life. I knew there were good and important reasons for me to stay exactly where I was at that moment: home with the kids, and available to Ma. And the writing was just helping me through; I cared about it the way one cares about food.

I looked at Pat, who was staring at his computer screen. For some reason I kept thinking about the time we'd visited a day care in Waltham when Zoë was fresh born. After talking to the director for a time, I had turned around to find Pat on the floor with three toddlers in his lap, reading a storybook, and I'd thought happily, *He knows how to do this, and he's on my team.*

I wanted that feeling back: *He's on my team.* I ached for some outside force that we could identify as our foe, something that would put us both on the same side again. I missed us-against-the-world, I missed Sybil and Pat. Now we felt like Sybil *or* Pat; somehow, every time my needs were met, he seemed to experience it not only as my gain but as his loss. I didn't trust myself to speak. I thought if I opened my mouth at that moment, I might emit a primal scream, channeling all the pain and joy of my life into one long earsplitting howl. How could I possibly make him understand this feeling inside of me? That was when Pat, perhaps taking my silence as some kind of a stubborn refusal and still not looking at me, quietly told me, "This is not what I signed up for."

Within a week we would seem normal again with each other; soon Cleo would go to day care, and eventually I would earn a living writing; much would change in the next year. But for the time being, that sentiment and those words lingered, like a scolding, like shame. I had no rejoinder, could provide him with no explanation or excuse to combat his broken expectations. All I could do was sit there, feeling inadequate and wrong, absorb his resentment, and stubbornly wait. Maybe something would change. I couldn't see what, but maybe something.

During that time I kept oscillating between two states: one moment I felt brand-new and full of joy, and the next I fell into a worried, pessimistic middle-aged funk. My girls infused me with happy energy, but the tension between me and Pat and my incessant worrying about Ma brought me down. In the face of certain alarming signals of my own aging process, I had to continually fight off the impression that I, too, was getting Alzheimer's.

One Thursday in February the sun emerged and sent all Berkeley into a heady preview of spring. On my way to meet a friend at the YWCA across from campus, I saw undergrads wearing shorts and tank tops sprawled on every lawn, soaking up the rare sunshine. I felt like the archetypal breeder, glowing and sexy and just bursting with juicy ripe ova. I wore a summer dress, feeling warm and sensual in the sweet breeze of this balmy afternoon. As I pushed the stroller lightly along the busy sidewalk, my body swayed lusciously of its own accord. Beautiful Cleo snoozed pinkly, a soft bundle of toddler plumpness.

I heard the jingle of keys behind me and glanced back to see two students: slim, brown, with smooth skin and large dark eyes. Their identical black bookbags were slung diagonally across their slim forms, causing perfect breasts clad in pastel knit to protrude just so.

"He's working for a big company now, doing telephone sales," one said.

"So he's a telemarketer?"

"Yeah! And I mean, this is, like, a career position, you know? I mean he works with, like, forty-year-olds; all his co-workers are, like, forty!"

"Weird."

"Good for me, though."

"Whadda you mean?"

"I mean, like, so he sits next to some forty-year-old woman all day? What could be more safe?"

And just like that, I felt the weight of my thighs.

A moment later I walked into the high-ceilinged, airy gallery where the YWCA invites local artists to display their work. As I waited for my friend, I noticed that one painting looked suspiciously like another piece I had seen printed many times on posters and greeting cards. In the more famous work, red, brown, and gold fabrics drape lusciously the length of the narrow painting, framing the forms of two lovers; her eyes are closed, and he bends to kiss her pale cheek. The painting in the gallery appeared to be just one swath of that fabric, at close range. I glanced around the room, trying to guess whether each of these abstracts might have been copied from odd corners of paintings by that artist. And I searched my memory for his name. Came up empty. I don't know if it was middle age, stress, sleep deprivation, or worry, but as Ma entered the middle stages of her illness, increasingly I found myself suffering from memory loss as well. Not Kandinsky. *K, K.*

A mild panic rose in me. Why couldn't I remember? I had always loved his work. I had owned prints of that kiss—that Kiss—for one long moment I struggled.

Klimt. Gustav Klimt. *The Kiss.* Mentally, I crouched, panting and sweating with the effort of retrieval. I felt the relief of having bridged a perilous gap. *But I've always done this, I* told myself. *Tip of the tongue. It's a common phenomenon. This is normal. This is perfectly normal. This. Doesn't. Mean. Anything.*

This was true. In a healthy brain, even after thorough learning, memories can dissipate with time unless frequently reinforced. Every time we revisit a memory, or retell a story, we reinforce the particular combination of emotion, sensation, thought, fact, and context that it is made of. At the cellular level we cut the memory path a little deeper each time we travel it, strengthening dendritic connections and thus increasing the tendency of that specific set of synapses to fire. But even when memories are reinforced, they may well be subtly changed during each retrieval. It has been shown that our current state of mind, the context in which we recall an event, can and will cause the details of a memory to change over time. In a way, it's silly to even think of a memory as a concrete entity, given that as living organisms, everything about us, including memory, is formed by a jumble of molecules that is constantly, subtly moving and changing. New enzymes replace old as we consume energy and expel waste, new cells are born and old die, the shape and size of us are always shifting; it is the nature of life itself that every part of us is continually in flux.

Compulsive worrying that one has Alzheimer's is extremely common among Alzheimer's caregivers. Like the medical student who begins to experience all the fearsome symptoms of the pathological conditions she is studying, we begin to take on the characteristics of our demented loved ones.

Unlike the medical student, however, I knew that I actually

had some legitimate cause for worry, because Alzheimer's disease does have a genetic component. About one hundred families worldwide carry what are called deterministic genes, which are deadly; if you inherit them, you are guaranteed to get the disease. Fortunately, fewer than five percent of Alzheimer's cases are caused by these genes, and those tend to be early-onset victims, showing their first symptoms in their thirties and forties. If such a gene were running in my family, Ma, Alice, and I would have certainly known of other family members who'd had Alzheimer's early in life.

However, there was still a chance that Ma, Alice, or I might carry risk genes, which are genes that simply increase one's chances of getting the disease. You don't have to carry such a gene to get Alzheimer's, and carrying it doesn't mean you will definitely get Alzheimer's, but statistically it is more likely. The best-known risk gene is called APOE, which stands for apolipoprotein-E.

APOE is a protein that carries cholesterol and lipids in and out of cells. Different people carry different subtypes of the APOE protein, and the combination of subtypes an individual inherits plays a role in the likelihood that she'll have the disease. We all get one copy of the gene for APOE from each of our parents, and that copy can come as one of three subtypes: e-2, e-3, or e-4. Subtype e-4 is the troublemaker. The other subtypes are harmless, but a copy of e-4 from one parent increases one's chances of getting Alzheimer's. Two copies, one from each parent, increases them even more. I constantly wondered how many copies of e-4 Ma had, and how many I had, and from whom. But I asked, and the doctors refused to perform genetic tests to answer these questions, because they predict so little: two copies of e-4 don't guarantee the disease will develop. It's possible to inherit two copies of e-4 and live to one hundred without any trace of Alzheimer's disease—or to inherit none at all and have the worst case of

Alzheimer's you can imagine. Other factors are clearly involved in determining whether we get the disease; we just don't know what they are.

Why does the e-4 subtype give some people trouble? One idea comes from the observation that people carrying that subtype have less APOE protein overall. Since APOE proteins carry cholesterol and lipids into and out of cells, it is likely that processes that require cholesterol and membrane (which requires lots of lipid) are disrupted in the e-4 patients.

One process in which cholesterol is essential is myelinization—the wrapping of axonal fibers in the myelin sheaths of glial cells to insulate our nerves and increase their speed of communication. Cholesterol is one of the key components of myelin.

Myelinization is crucial in early neural development, but myelin sheaths also continue to accumulate until fairly late in life. The process is thought to be complete around the age of fifty, at which point demyelinization—the unsheathing of nerve fibers due to myelin degradation, which is a natural consequence of aging—begins. So glial cells surrounding the neurons in Zoë and Cleo's brains and even in mine were still in the process of wrapping themselves tightly around our smaller nerve fibers, just as Ma's myelin began to come unwound. But in Alzheimer's and other neurodegenerative diseases, demyelinization begins earlier and proceeds more rapidly. It is possible that people carrying the APOE e-4 subtype, because they have less APOE protein overall, have some imbalance or shortfall of cholesterol that leads to a problem with myelin, maybe causing it to break down more quickly. If Ma was one of these carriers, her sheaths may have been unwinding faster than was normal for her age.

Lipids, the other type of molecule carried by APOE proteins, are an essential ingredient in the making of acetylcholine (ACh, the neurotransmitter that goes missing in Alzheimer's disease).

Since lower than normal levels of APOE in the hippocampus and temporal cortex of Alzheimer's patients with the e-4 subtype may cause them to have trouble with lipid transport in the brain, this in turn may interfere with ACh production.

And finally, when APOE is mixed with amyloid in the lab, it seems to slow the formation of amyloid plaques. Thus, whatever the cause of the plaques may be, carriers of the e-4 subtype, who have a shortage of APOE, could be at a disadvantage in staving them off.

There are other risk factors for Alzheimer's disease. Several factors affecting the cardiovascular system together can make a person three times as likely to get the disease. These include diabetes, hypertension, heart disease, and current smoking. Additional risk factors include low income, a low level of education, depression, prenatal deprivation, and finally, a big one: head injury. *I don't have those,* I told myself ten times a day, *I'll be okay.* This didn't help a bit. To my knowledge, Ma had none of them, either. Besides, I knew that the biggest risk factor of all was age. I was going to get Alzheimer's, if I didn't have it already; I just knew it, because no matter what I did, the symptoms kept showing up.

One afternoon when I picked Zoë up at kindergarten, her teacher, Ms. Rosen, had just finished a unit on quilts, and I picked up ours to bring home. It was a beautiful piece of work, plush deep blues and purples. Our friend Heather had sewn it by hand for Zoë when she was born. But later, at home, I could not find it. *Wait,* I thought, *where did I put it?* And then: *Did I really bring it home, or did I forget?*

The next morning I lost the grocery list. I was sure I'd had it in my back pocket, but when I reached for it, it was gone. I wandered around the house looking. In the kitchen, in my stack of greasy, scribbled recipes? On my desk with notes from voice mail? By the bed, under my list of things to do today? I could not find it.

After several minutes of searching, this no longer felt like what my friend Sophia called "Mom Brain." It felt much more serious to me. There I was, shuffling through the piles of paper, looking for a list. Why couldn't I just make a new grocery list? Because I would forget something. *That's it,* I thought, *this is how it starts.* I tried to take deep breaths, but I couldn't seem to get enough air. *I have it, I have Alzheimer's, oh God.* My throat was closing up, and I felt the tears burning in my eyes as I did one more sweep of the living room, overturning pillows, as if I might have secreted my list away in the couch. *I have to find that stupid thing,* I thought. *How could I have lost it? Stupid, stupid.*

Once again I was starting to resemble Ma. This time, though, I wasn't alone on the street with my worries, I was in a house full of family. The girls were in the kitchen eating Honey Oh's cereal when Pat heard me sighing and sniffling in the living room.

"Hey, what's going on?" he asked, concerned.

I could not look at him. I stifled a sob, and in a raspy whisper I told him, "I lost Zoë's beautiful quilt, I can't find the grocery list, and I can't even remember what was on it. My whole life is in these little lists of things to do, and now I'm losing the lists!" Here my voice jumped an octave. "And I'm wandering around looking for them in all my little piles of *lists*! And—and it's just like her! I'm turning into her! I think I *have* it!" I sobbed some more, against his chest.

"Mama?" came Zoë's quiet voice.

"Oh." I wiped my eyes and then peeked over Pat's shoulder to see her standing there, looking up at me. Her blond hair was sticking out every which way, and her dark little eyebrows pulled together in a bunch.

"What's wrong, Mama?"

"I'm—I'm confused," I said. "I forgot some things." I was thinking: *I'm going to forget you. I'm going to look like me, but I'm not going to act like me. I'm not going to be me, for you. Oh, Zoë.*

Zoë reached up her arm and encircled my waist. We all sat down on the sofa together. "Why do you think you are forgetting?" she asked.

"I'm not sure." That was as close to the truth as I could go; I wasn't ready to tell my five-year-old, *I'm afraid I'll get Alzheimer's disease like Gram and slowly lose my mind.*

I had tried to do all the things I was advised to ward it off: get plenty of physical exercise, mental exercise, and social interaction; try not to get too stressed; maintain a healthy diet, including plenty of turmeric (Indians, who eat plenty of curry, have a lower risk rate) and vitamin E; and don't drink excessively, since that might cause a dementia of its own. Some studies hinted that a moderate glass of red wine with dinner and maybe a small dose of cannabis for dessert might help, but I had decided against those, since pot gave me, of all things, memory problems, and alcohol made me depressed. I already felt sad and confused, and I had read that people with a history of depression were more likely to decline rapidly once they had Alzheimer's; they had more plaques and tangles in their hippocampi than happy folks. You just couldn't win.

"Have you had breakfast yet?" Zoë inquired.

"No, not yet."

"Well, I could suggest that you eat breakfast. That sometimes helps me."

I blinked away more tears and turned to hug her, smiling at Pat over her shoulder. I tried to let that be the end of it, but I felt like a loser, a frightened monster.

Then, just as we were about to walk to school, Zoë went

upstairs with Pat and would not come down. I called to her weakly. I didn't have the energy for the standard get-to-school battle.

She yelled down, "Just a minute! I'm having Daddy do something important for you!"

A minute later she tromped down the stairs and handed me a handwritten note. "Here you go," she said, with the confidence of a physician handing me my prescription. In Pat's handwriting, it said:

Zoë asked me to write this down for you:

When you are unhappy,
Your brain is unhappy,
And when it's unhappy,
It doesn't remember well.

Zoë. She was better than a therapist, as supportive as any support group. I tucked the little note in the front pocket of my backpack. If I forgot, it would be there to remind me.

CHAPTER 19

Lost and Found

"Ma? Do you know where the old Halloween stuff is?"
Zoë and Cleo are in school and day care, and I have arrived early in Martinez to take Ma to her annual Alzheimer's checkup. Yesterday I thought of a way to get out of making Zoë's costume this year, and Ma doesn't know it, but she's going to help me.

"We used to keep them up in the bedroom closet," Ma ventures.

Yeah, about a million years ago. I root through every corner of my childhood house, unearthing old toys, stuffed animals, and dress-ups. Finally I dig the box, labeled in her neat calligraphy, out of the basement.

My breath catches as I lift out the clown costume Ma sewed for me when I was five years old. The baggy suit is so familiar and so perfect: half red, half white, with three yarn tassels, a ruffled collar, and peaked cap. I loved it so much, I wore it three Halloweens in a row. It will just fit six-year-old Zoë. I examine Ma's perfect seams. I can picture her, up late, bent over the sewing machine where she made her own dresses, Daddy's vests, our costumes. That machine is at my house now; she has forgotten how to use it. In my head, though, she's still the expert. I'm only marginally competent with the thing. The bobbin always pops out; I break needles and swear under my breath, while Zoë stands nearby, anxious, perplexed; eventually I shrug and hand her a weirdly shaped article of clothing with half the seams inside out. This clown suit is worlds better than anything I could produce.

After unearthing the costume, I drive Ma to the Alzheimer's Center. This will be the second checkup she's had since she was diagnosed just under three years ago. We follow our usual routine: Ma, grumpy, acts like she's doing this for me and grudgingly submits herself to memory tests. I try to see myself as a compassionate health advocate rather than as a bossy daughter, and I end up asking too many questions. Then Nurse Carol tells me privately what I already know:

"Ruth's condition has worsened." She says these words carefully, looking into my eyes with warm concern. Then, more firmly, she warns, "You should find her a professional care facility. Soon." I glance over her shoulder at the puppy dog calendar on the wall behind her. Miniature American flags run along the border of the photo: four chocolate Labrador puppies asleep on a green velour couch. *Two thousand and three,* I think ruefully, *the year I moved Ma to a nursing home.*

I ache to be alone, to cry. As we leave the building, Ma stumbles. She looks haggard, meek; she's walking funny, and I have the impulse to pick her up and carry her to the car like a child.

Afterward we lunch at her favorite diner, Peg's, which has a view of the Tosco oil refinery with its bizarre jumble of tanks, tubes, vents, and smokestacks stark and gray against the blue sky. Ma is silent; she looks exhausted. I'm sad, overwhelmed.

"What do I like here?" she asks.

"BLT," I answer automatically.

I see another question forming. "Bacon, lettuce, and tomato sandwich," I say. "And you usually like a small orange juice."

I watch as she struggles to chew her food at lunch, with her mouth open and her head drooping. Halfway through lunch, she begins to hiccup and doesn't stop. This is the worst I have ever seen her.

After lunch I take Pine Street home, because I like its sweeping view of the Carquinez Straits, and because it takes us past my old junior high school. The brown hills rising up behind the school are so deeply familiar, my heart seizes up. I revisit a feeling I had in the fall of eighth grade: in the first days back at school, those hills were still the golden brown of California summer. That struck me as somehow dissonant, somehow tragic. There we were in our stiff new back-to-school clothes, backpacks heavy with fresh textbooks, intent on our schedule. Yet those hills, browned by a blazing sun, had not moved. The day, clear and bright, meant business, but the hills just sat there, a living reminder of the slow, easy heat of summer. The crickets still sang at night, and the fire hazard was higher than on the Fourth of July. The hills looked down and said, *This school is just artifice; the sexy daydream days, comic books, long bike rides, hours lolling in the pool—they all live on, in this brown dry grass.* I was stuck in that harsh house of rules where bells rang and I rushed to class; I had to do homework, had to wear a bra, had to act right, had to had to. But the hills didn't care. They just sat, round and golden, peaceful, prickling in the bright sunlight, breathing out the last breath of the season.

I feel a similar dissonance now when I am with Ma. This is a new season for us. I can no longer reach her; we can't really talk. She no longer thinks like Ma; she doesn't initiate, articulate, or enjoy like Ma. She doesn't care. Now everything has shifted, and she's not the Ma I knew. Yet she's right there in front of me, looking for all the world like herself. I cannot deny that it is she. Her face, especially her smile and laugh, say, *This is ridiculous—here I am. I have been here all summer—nothing has changed!* If I just had to take care of any old bewildered elder for a few years, I wouldn't complain. It's Ma's physical presence, the fact that there *she* is, that makes me crazy.

I find myself searching for proof of the mom I grew up with— like that old clown costume, so expertly sewn. I think of the way physicists study subatomic particles using a cloud chamber. The particles cannot be directly observed, but whenever one moves through the chamber, water condenses in its path, creating a vapor trail. The physicist eagerly photographs this evidence of the moving particle, before it disappears. The clown costume is like that elusive trail, the beautiful colors and neat seams tracing one path that Ma's intelligence took, decades ago. Yet it can no sooner capture the truth and magnificence of her warm, merry, eccentric self than the physicist can catch a neutrino in a wooden box.

Ma hiccups the whole way home. As we turn onto the street I grew up on, my mind shifts back and forth between childhood memories—the troop of kids playing hide-and-seek at dusk—and my adult perspective, wondering how much our funky old house is worth now. We'll need every penny for extended professional care.

By the time we arrive at Ma's house, I have one hour until pickup time at day care, so I pull up to the curb at the house and wait in the car as she steps cautiously across her dead lawn, stops to check for mail, and bends to unlock her front door—my old front door. I hear the familiar squeak of the screen door hinge. From the back, in her wrinkled trousers, work shirt, and stained old hat, Ma looks just like Gramps, her dad.

She opens the door, and I see the darkness inside. She never opens the shades anymore. Will she look around at this home, one of the last familiar places left to her, and take comfort? She must be tired after her appointment. Was it traumatic? Is that why she got the hiccups? I picture her sinking into a chair, sighing with relief at the intimate feel of it.

Sitting in the car, watching the door close behind her, I picture it, but I don't know. It's a secret. I don't know what she does there when I'm gone, I don't know what she fears, how she feels. She's never talked much about her feelings, and these days she doesn't talk much about anything. It's frustrating, but I remind myself that there's a disease inside her head, gluing her thoughts together. Even if she wanted to tell me, she might get stuck, trying to find the words, trying to capture elusive emotions; she might forget what it was she had been feeling. Maybe she doesn't experience emotion the way she once did. I don't know. Inside the dark house, with the doors closed and the shades down, her small, private life is utterly opaque to me.

I don't get out to see her in. She doesn't turn to wave goodbye.

That night, back in Berkeley, Zoë is resplendent in her clown costume. She hops around before the mirror, jiggling her tassels. She can't wait the three weeks until Halloween. "This is *so* much better than waiting for you to sew me one!" she says. I smirk at her, feigning offense. But then I have to smile, because here is Zoë in a suit tailored by her grandma—or, more accurately, by my mom. And there I am, five years old, beaming up at Ma, loving her.

When the costume is off and Zoë has settled in with her first grade homework, I put on broth for chicken soup. The house is unusually quiet for six p.m., with Zoë scratching away at math problems and Cleo in the living room whispering to the small plastic dolls she calls her "people." I wonder if I have time to sneak in a phone call to the home we've chosen, to make sure we're still on the list and inquire about the next opening. I remember Ma's words: "If I have to move from here, I think I'll just run away . . . to hell." Still, while I can't stand the thought of moving her, Carol is right; she needs more help than I can provide.

The cleaning lady I was going to hire to check up on her didn't work out; Ma met her and disapproved. "I don't like it," she said. "She's a stranger, and I don't feel comfortable with her. Besides, what would she even do here? Please tell her to go away."

Ma has also been resisting my help cleaning, which is odd because for many months she barely seemed to notice me doing it. Now suddenly it's an offense, a disruption, and it agitates her. Last week I confronted her about the grime that had been accumulating in her house. I pointed at the kitchen floor.

"It's almost black, Ma! I'm going to mop it."

"Don't you do that! I can clean my own floor."

"Ma, it's filthy."

"I don't know why. I clean it."

"How do you clean it?"

With a dirty damp rag, she swiped at a patch of floor. "See? Nothing more comes off."

I found an ancient bottle of cleanser under the sink, poured a little out, and rubbed. A bright beaming circle of Formica emerged and sparkled up at us accusingly.

After that I started looking for another cleaning lady, but now it looks like that won't be necessary. Nurse Carol has made it perfectly clear that Ma needs more than a house cleaner. Of course I knew this, but I'm feeling relieved to have such clear marching orders.

In the end, though, I don't call the nursing home. Instead, I call Ma to check on her myself. After today's memory testing, she was so bedraggled and inarticulate, I think maybe she's coming down with the flu. I want to see if her hiccups stopped and make sure she got dinner okay.

After five rings I get the answering machine. It's late for her to be out feeding the ducks. Maybe she's in the yard talking to the neighbors with the puppy. Or could she still be napping? I imagine her in bed, fully clothed and softly snoring, and the thought fills me with melancholy. She lives alone. No one expects her for dinner. In my living room soon the girls will be yelling and pushing each other, hungry and impatient. I could never sleep through dinner. Thinking of Ma there alone, I feel a rush of gratitude for these needy little beings in my life.

By the time Pat comes home, it is seven-thirty. The girls and I have already eaten, and I have tried Ma's number twice more. Zoë and Cleo parade around the kitchen, playing the tambourine and giggling while I wash dishes.

I raise my voice to be heard over them. "My mom's not answering the phone."

"What?"

"I dropped Ma off after lunch. She should be eating dinner, but there's no answer."

As I say it, I realize how wrong that feels. Ma goes by the clock; she eats every day at exactly six p.m. If she did nap through dinnertime tonight, it would be the first time, ever.

I think of the question Nurse Carol always asks: "Has Ruth begun to wander at all?" I picture Ma hurrying along a highway shoulder, wide-eyed and disheveled, as cars whoosh past without slowing.

"I think I have to drive out there," I say. "Can you handle the kids if I go?"

Patrick glances at Zoë and Cleo, and I can see the objection coming; he hates to be left with the two of them when they are tired.

"Please?"

"I'll go," Patrick offers. He'd rather save the day.

I pause while something clicks into place inside my head.

"No, I think I have to go," I hear myself say. It has to be me. There's a rising excitement, a strange burning, that tells me this. And then I want a drink. Something strong, to match the urgency mounting inside me. It's dark outside. She should be awake by now. What if she's not there . . . what if . . . ?

A familiar mixture of compassion and annoyance washes over me. I hate to leave the kids and the comfort of home right at bedtime, but there's no question that I need to check up on Ma. I grit my teeth, and then heave a sigh of resignation.

On some level I'm aware that I've been waiting for a crisis— because a serious enough crisis is what it takes to finally land a person in a nursing home. I might really be on the verge of turning her care over to a group of perky and efficient paid attendants, who will greet me with a progress report when I visit every Saturday morning. This is both my fear and my fantasy.

I kiss the kids at the door. At two, Cleo is agreeable, but six-year-old Zoë has already absorbed my distress. I struggle to reassure her without lying. I want her to feel my concern for Gram more than my fear of what I might find. In Zoë's young face I see alarm and love, and then determined, mature acceptance.

"Okay," she says, "I understand. You have to make sure Gram is okay. You'll call us, though, right?"

"I will," I say firmly, and I look from Zoë's face to Pat's and back again.

Ma's house in Martinez is a forty-minute drive from mine in Berkeley. It's warm outside. I drive onto the freeway with the windows wide and the stereo blasting. I surge with a rare but familiar energy: freedom. Just like Ma once did, I relish driving. I'm out at night, by myself, and the possibilities are endless.

With every mile, the desire to drink hard liquor grows. With it comes the fantasy that I will just blow through Martinez and keep going; I will speed all the way to New York, where it is cold and dark, and I have no responsibilities. I will drink myself into oblivion.

But soon I'm thinking of my girls, flushed and drowsy and cuddled together in their bed by now, and soon I'm aware of the electric current in my chest that draws me to Ma. I know that if I did run away to be free of all this, I would immediately feel bereft. The feeling of captivity that comes from Ma's constant neediness, the work and sorrow and annoyance of it, also keeps me fully present in my life. I go to Ma because she makes me whole, as she always has; I hesitate to leave my girls for the same reason. I'm hooked in. Making that trip from my daughters to my mother, I connect us all to one another, complete the circuit between our three generations. This is my path. The illusion that I might find freedom in escape dissolves, and I travel the familiar road to my old family home.

When I pull up, it's almost nine p.m. I've just hurtled twenty-five miles through the night, and one part of me is filled with dread, while another wants to kick myself for worrying so much. There are no lights on, but the TV flickers behind the blind: I sigh with relief. Still, she doesn't answer the door, and there's a faint hissing sound I can't identify.

I knock loudly and call to her. Then, feeling like a thief, I use my key.

I follow the voice of the television through the dark house to the dining room.

"Ma?"

Light plays on empty chairs.

I recognize the hissing sound now; the kitchen faucet is on full blast, in the dark. I reach around the corner for the light switch, which my hand knows by feel, and warm yellow light fills the kitchen.

There's Ma, lying on her side, on the dirty Formica floor. A wide, thin sheet of brown vomit has spread out around her. Her blue eyes are open. She's very still.

From the dining room, music suddenly swells, melodious and poignant: layers of celestial female voices, loving, gentle, strong. My eyes fill with tears, even as my critical mind marvels that such appropriate background music should rise up from the television at this precise instant.

I look at Ma. Is she dead? I think she's dead. I should call someone . . . I start back into the dining room, for the phone, and there's the television again.

My mind is quick but not smart. I'm moving from fact to fact, and for a moment each data point takes on equal weight; I can't distinguish the vital information from the expendable. For several seconds I stand paralyzed, halfway between the big, bright color TV and the body of my mother. Then I return to her.

She's lying on her left side. Her hand is curled up by her face, as though she's just brushed away a stray hair. Her open eyes keep insisting that she's there, even as they inform me that she's gone. Her expression is attentive, without drama: she is waiting for her coffee cup to be refilled, for the librarian to check her books. But the tip of her tongue has a strange curl to it, and her fingernails are blue. I reach for her wrist, which is cold, and feel for a pulse.

"Ma?" I say. "Ma? Oh, Ma, I think you're dead."

Then I finally walk to the phone and punch in my own number.

"Patrick?" I breathe. "I think she's dead."

"Oh, Sybil."

"I'm not sure, but—no, I am sure. But what if I'm wrong? I can't think what to do. What do I do?"

"You can call nine-one-one. Someone will come. They'll make sure." There is no distance between us; Pat makes himself fully present for me. His voice is calm but full of compassion. Months of tension, resentment, and complex negotiation fall away in an instant, and all I feel from him is the pure essence of his support and love for me. "Call them now, and then call me right back," he urges, and he's right, and my balance and reason return to me in a flash. I'm so glad I called him.

When I phone him back after calling the police, he immediately offers to drive out. "I can call someone to watch the kids," he says. "I can be there in forty minutes."

"No, it's okay. I think I'm okay here."

"I really feel like someone should be with you."

"It's late. You'd probably have to wake somebody up."

I would call Daniel, who lives just across town, but he is in Oregon at a Shakespeare festival this week. I'll call him in the morning. I wish I could talk to him now. He's the closest thing to a father I have left. But why am I telling Pat not to come? I shut my eyes and try to identify the panicky feeling I'm having. And I find this: I need Pat to be with the girls. I need him to go in and check on them and make sure they are okay—I need to know he is with them. Right now, with this happening here, somehow I can't stand the thought of my babies being away from their parents, or of Pat hurtling too fast down the dark highway alone. I know it's not rational, but somehow I need them all safe in our house, together.

I'm trying to explain this to Pat when I hear the siren. I imagine a car wreck somewhere in the night, but then I realize someone is coming to take care of Ma—and me. I hang up to greet the firefighters, and I feel so grateful for the huge red truck, which fills the entire narrow street, quietly huffing. The firefighters and police officers fill the house with their big bodies, glossy equipment, and gentle, clumsy assurances. Their strange smells and big male faces feel fatherly to me. I am small and dazzled, like two-year-old Cleo on a Tuesday morning, waving to the garbage collectors.

"Ma'am, I'm very sorry for your loss. Ms. Lockhart, is it? You're the daughter? Can we just ask you a few questions, here?"

Suddenly I'm an adult again. I'm not crying. Do I seem cold to this kind, burly man? But there is nothing I can do about that. I feel myself calmly stepping in to oversee Ma's death; the habit of managing her affairs is so strong. A police officer with a notepad asks me how old Ma is and whether she was ill. When I say she had Alzheimer's, he nods, as if that's enough of an explanation for him. The firefighters move into the kitchen, and as I smoothly deliver facts to the officer, Ma is redefined as The Body.

Too soon the fire truck has pulled away. The coroner still hasn't approved release of the body to the mortuary, but the police officers tell me they have to leave, too. There's a lot of methamphetamine around, they say, and it's Friday night. Their supervisor wants them back out there.

When they go, I have an irrational urge to leave the front door wide-open. Being closed inside the house with dead Ma doesn't feel right. *But I'm a biologist,* I remind myself, *and this is just a dead body: the most harmless human possible.* Better to be locked inside with my sweet mother's body than to leave the house wide-open at midnight in a town in the midst of a meth epidemic. So I close the door.

"Ma's dead," I say out loud. My voice sounds false and wavery. I feel claustrophobic and panicked. Until I go to Ma. I don't want to go in there, yet I need to. I lean in close and rest my hand on her solid waist.

"Ma, I am so sorry you died."

I want her to feel safe.

"I love you, Ma."

I try out the past tense: "I loved you."

This is all wrong. For years now I have been planning for our future—a dreary future, but a future nevertheless. Ma was to go slowly, lose ever more of herself, be moved to a home where she would be nursed and coddled as she grew older. I would visit her every week, sit by her bedside, and comb her hair after she had forgotten my name and lost control of her bodily functions. Now that path, which just this morning snaked out a decade or more into our future, is suddenly closed. Already I am beginning to obsess about how and why this happened, as if knowing the answer could possibly give me any measure of control. I go back over the afternoon in my mind, and the symptoms that I interpreted as the flu—the burping at lunch and the sudden fatigue—no longer seem flu-like to me. Ma's incoherence and her odd gait, which just hours earlier I took as signs that the Alzheimer's disease was progressing, suddenly look more like a stroke to me.

She must have been rinsing the dinner dishes. What did she feel as she turned from the sink, with the water still rushing from the tap, just before she died? I remember in fifth grade how my friends and I would deliberately hyperventilate and then hold our breath in order to pass out. I would fall onto the grass one moment and wake from a faraway crazy dream the next to find the others laughing over me. Did Ma experience that same lightheaded oblivion, but just never come back?

As I scan Ma's hair and face, her still limbs and open eyes, I know that some life remains in her body: there are muscles that might twitch, and cells that could be coaxed to survive and multiply even now. I know that I will never be able to speak to her again. I know that legally she is dead because her heart, her lungs, and the core of her brain have all irreversibly ceased to function. But the process of death in the rest of the body is long and nebulous, and my heart tells me she isn't entirely gone, no matter what the fire department has decreed. I think about the second wave of destruction in the brains of stroke victims, a suicide signal sent out to the cells neighboring those that lost oxygen and died in the first round. Science's more evocative name for programmed cell death rings weirdly in my mind: *apoptosis.* It's a strange word. Because of the *pop* part, I always imagine the dying cells popping like little balloons. In fact, the word comes from the Greek *apo,* "from," and *ptosis,* "falling," as in leaves falling from a tree. Glancing at the cans and pot lids strewn on the floor around her, I wonder if Ma fell, scattering them on the way down, or if she knocked them down and then lay down beside them.

For the next two hours I wander around the house, crying and staring at things. I call Pat again, and he keeps me company for a while; I call Alice, who seems more than anything sorry that I have to be there without her; but later I won't remember the details of the conversations. Nothing seems solid. I keep returning to Ma to say a few words, trying to memorize her face. I say, "I hope you were okay. I hope you didn't feel scared or lonely when you died. I love you. I hope you knew I loved you."

When the mortuary men finally come, it's one-thirty in the morning, and I still haven't cleaned her up. I've been too busy having my feelings and this one-way conversation. I think, *It must seem strange to them that I didn't even wipe off her face or close her eyes.* Embarrassed, I distract myself by asking them questions.

"Do you guys always dress so well? Do you have any spare gloves? I've never seen such a big plastic bag. Do you suppose it's made by Glad?"

After they are gone, I am stricken. They have left her glasses on the piano bench, next to where the gurney stood. I can't look at them. Suddenly, I miss her unbearably.

I turn back to the kitchen and slowly go about cleaning up the enormous pool of vomit. *Why vomit?* I wonder. *Does that often happen just before a person dies? And why did she die?* The mess is not horrible or smelly, just icky. Years of stinky diaper cleanup have prepared me for this moment. I find rags in the kitchen drawer, and I swab up the dark liquid as best I can, pouring on some of the old cleanser from under the sink and trying not to inhale, in case she really did have the flu.

When I'm finished, the entire floor is a much lighter shade of brown, but I can still see the clean spot I rubbed away when we argued about her diminished housekeeping skills last month. I won't have to hire Ma a helper; I won't ever have to call the home. This thought, more than any I've had this evening, undoes me. It is the first responsibility to be lifted, the first "last time" I will have to attend to a caretaking task. What I once thought would come as a relief leaves me empty and alone; the world has just expanded around me, and I am lost.

CHAPTER 20

Days of the Dead

Sometimes the mundane details of parenthood can seem downright poetic. For example, I routinely found a stray Cheerio or two on the floor after breakfast, and it was my habit to stoop and retrieve these escapees as I made my way through the room. Every now and then, though, a funny thing happened: I would encounter a Cheerio that looked normal but had in fact been stepped on. I called them Stealth-ios, because at first they didn't appear crushed; they still held their perfect little round Cheerio shape. But if I tried to pick them up, the instant my finger touched them, they disintegrated into little piles of pulverized Cheerio powder. Now that was me. *Just try it,* I felt like saying. *Try and touch me.*

The night Ma died, during the hours I waited with her body, I stayed close to her, filling myself with the knowledge of her passing. When the mortuary men lifted her onto the gurney, I heard a small snap, like the breaking of a brittle twig, before they fed her body into that enormous plastic trash bag and drove it away: I confronted the physical reality of her death head-on.

Halloween, only weeks later, was impossible. I felt adrift, and the images of death infected me. As I drove past the temporary Halloween store, I saw the anemic flesh and felt the open-eyed stare of the fake corpses in a new and personal way. The mock graves in my neighbors' yards made me shiver and cry. Every time I let my mind rest, I saw Ma's face in death, those blue eyes open and waiting, the tip of her tongue just visible and perfectly motionless. The image didn't frighten me, but it reached deep inside me every time and shook me, hard.

As I knelt next to her, speaking to—what? to her, or to that part of myself that still carried her?—I felt such an intimacy, not only with Ma but with the very oldest parts of myself as her daughter. There she was and there she wasn't, in front of me but no longer there. As I watched her inert body, I felt that the weight of the two of us was being redistributed, consolidated; in a way, I was taking her in, taking her on. I would carry her forward, and so would Alice and Zoë and Cleo, because in so many ways she was still inside of us.

On another level, she was just gone. I discovered that no matter how diminished a person with Alzheimer's disease may seem to her family, she was infinitely more alive *alive* than she was dead. Dead people didn't do a thing. They didn't lose their lists. They didn't fret or hover. They didn't expect you to call every day. They didn't read to their granddaughters in an odd, halting cadence. They didn't even, on occasion, gently laugh in a way that reminded you of happier times.

When I found her, the stillness of her body transfixed me. I didn't need more information; it was all there in the kitchen, more accurate, more complete than any pathologist's report. Now, though, the habit of caregiving persisted; I still worried about Ma. I no longer fretted about where she would live or whether she was safe, but I wondered obsessively about the moment of her death. In the inverse of the urge I had once had to swim among Zoë's burgeoning cells in utero, I yearned to follow Ma, just far enough to know that she had arrived somewhere safely. Every morning I was struck anew by the agony of not knowing. Was she lonely? Had she suffered?

And why had she died so suddenly, anyway? The detective story of precisely which disease she had had dropped away and was quickly replaced by this new mystery, as I continued my futile scramble for control over uncontrollable situations. Ma left us on the cusp of the most harrowing stage of her disease, just in time to spare us the slow, clinical hush of the nursing home. The coroner never ordered an autopsy, so there was no known cause of death; the undertakers finally had to come up with a condition common among her demographic to list on her death certificate: athero-sclerosis, a hardening, narrowing, and blocking of the arteries. But that was just a guess. I fretted for weeks about what really killed my mother. Why had her body shut down so fast? Could she have overdosed on her Aricept the day of her psychometric testing, in a childish attempt to ace the exam? Did that cause the stroke I had come to so strongly suspect?

A part of me felt embarrassed that I hadn't ordered an au-topsy; I was a scientist, after all. Really, though, it didn't matter. I knew that if Ma could have made a conscious decision to die at this juncture, she would have. She would have yearned to spare her daughters the heartbreak of tending to her through the late stages of this horrible disease, and here she managed to leave us the very day I'd been ordered to move her from her home. Maybe, in the end, death was just the decision she came to that afternoon. Though my worries kept swelling up in waves, as each wave sub-sided, I returned again to the flat reality of death: I had no power over this. No additional information would change that one fact, nothing more mattered; there was nothing to *do* about death.

Not that there wasn't plenty to do. Ma had named me executor of her will. When I had first become aware of this, it had
surprised me. I had always been the irresponsible one, impulsive,
adventurous. Alice was older, dependable, a manager; to me, she
seemed the obvious candidate for anything involving money and
lawyers. I was flattered to find that long before I had returned
to California, Ma had placed her trust in me. She had extended
her faith. Now, though, I had to untangle her finances, empty her
house, sell her land, and distribute her savings, which, amazingly,
would no longer be sucked into the vacuum of assisted care but
instead would be equally divided between Alice and me.

We decided to put off a memorial until December, when
Alice's whole family could come down from Seattle. Ma had
prearranged cremation of her body, and I would go to collect
the urn when this was done. In the meantime, Alice flew down
alone, and in one whirlwind three-day weekend, we pillaged Ma's
house. Alice ran through making piles of her old stuffed animals
and toys, the odd item of costume jewelry, dishes, and vases, and
many, many boxes of books. I tried to do the same. It was tire-
some, dusty work, and at every turn there was another memory
to confront, from Daddy's old toolbox in the cupboard next to
the small brick fireplace to the dozens of beaded necklaces I had
made Ma for birthdays and Christmases through the years, which
hung in the dark bedroom on a rotating rack that Daddy had
created from plywood and metal coat hangers. These particulars,
which had always been unremarkable fixtures in the house, suddenly became icons of our childhood and our parents' ways; I
felt myself getting drawn in by a powerful urge to preserve everything. So much of what was there was unsalvageable junk, yet it
was *ours,* and I didn't want to forget.

Odd, unfamiliar items surfaced. On a high shelf I found a set
of dentures. No one in the family had ever worn dentures. They

had gleaming pink gums, authentically yellowed teeth, and an encrustation of sea barnacles all the way around. Who found them, when, who lost them, and how, were mysteries no one would ever solve. So much is either unknowable or forgotten in one generation, it seems futile at times to seek information about the past at all. The dentures may have been mere garbage from a barge, washed ashore and kept as a curiosity by my parents, or a clue to the circumstances of someone's dramatic death at sea, but now, I thought as I tossed them into one of the heavy-duty black garbage bags we had stationed in each room, the facts and artifacts of some stranger's life and maybe death would be lost forever in our town landfill. When I thought of that, I felt like less than a speck.

While Alice looked for things to keep, I went through the items I felt I must immediately dispose of: clothes, shoes, dog leashes, bedding. Somehow I had an urgent need to purge these items; I sensed I'd never be able to return alone and continue the job of emptying the house if they remained. I would sink down into the musty mess and be swallowed, I would suffocate on the past. Someone had told me the week before that I should systematically check the pockets of all the clothing, and when I did, I found tissues and tissues and tissues . . . and a twenty- and a ten- and a five-dollar bill.

"You want some of this?" I asked Alice, holding out the mound of crumpled tissues with a grin.

"Finders keepers," she replied.

"How about this?" I held out the bills.

"Whoa! Well . . . finders keepers, too, I guess."

I stuffed the bills into my front pocket. "How about if Ma buys us dinner tonight?"

"Deal."

Pat took the kids part of the time and helped package and mail Alice's boxes. Zoë romped around playing with our old toys, and Cleo got very dirty toddling and falling among the boxes. I went through bursts of activity followed by moments of feeling lost and empty. I was finding notes all over the house with my name on them: *Sybil's on Friday. Ask Sybil about shopping. Take to Sybil's house.* I tiptoed into the dark study and listened to my own voice on the answering machine: "Ma, it's Sybil. Are you there? I'm worried about you. I think I'll just drive out and check on you. Call me if you get this, okay?" I pressed rewind and listened again. I was so grateful to hear my reassuring, compassionate tone. How had I managed to sound like that in the state I'd been in? Had she heard the message play live as she lay on the floor dying? Could that be why she had that expression of patient waiting on her face; had she been waiting for me?

After the beep I sat still in the dark, silent room, feeling weak and helpless. I kept recalling the image of her as I had last seen her, stooping slightly to unlock her door as I watched from the car, my mind already back in Berkeley. I had to face the thought that I had been fighting off all day: What if I hadn't rushed off to pick Zoë up at day care that afternoon? What if I had been able to stay with Ma to feed her a snack and tuck her into bed? Maybe I would have realized something was seriously wrong, that she wasn't merely exhausted. I knew there was no use thinking such thoughts now, but I had been there, I had been *right there,* that afternoon. What if all my responsible acts—the diagnosis, the testing, the stress of it all—had driven her to her death? And that day I just turned around and drove away, leaving her just as the first symptoms of a possible stroke appeared, leaving her to fall. I couldn't stop that thought from coming. It continued to grow, filling me with remorse and shame. I couldn't stop it, so I simply picked it up and took it with me as I went back to work.

Sorting through random household goods, I found multiple copies of certain items: five bottles of glue, two knife sharpeners, four whetstones, three compasses, three magnifying glasses, no fewer than four pairs of binoculars, rolls and rolls of tape, multiple boxes of staples and paper clips, dozens of AAA maps, and a kitchen drawer full of lightbulbs. *Locate it,* these things said to me, *find it, illuminate it, hone it, keep it together, magnify it, fasten it, maintain its edge.* Maybe this just happens in any home where the same person has lived for decades upon decades: certain household items just accumulate. But the compasses? The binoculars? In a very tangible sense, these collections represented the struggle against change and blurry edges, the struggle not to be lost, to hold everything in place. I imagined her returning again and again to the hardware store or the stationery shop, subconsciously searching for something to fix things, to make herself feel safe.

Before closing the boxes, Alice and I peered at each other's piles. There was nothing to contest; we were in complete agreement about who got what, with the exception of one slim black volume of T. S. Eliot, *Murder in the Cathedral,* with our father's neat calligraphic signature on the inside cover:

Robert P. Lockhart
Reed College
Sept. 1950.

"Oh, you take it." She shrugged after about ten seconds of tension. "I have so much here, and I'm leaving you with everything to do." *Don't!* I wanted to say. *Don't leave—I'll give you the Eliot, just stay here with me.*

She couldn't stay. Early Monday morning she was on a plane back to Seattle. I dropped Zoë at preschool, and Cleo at the little home day care where she had stayed during my lectures, and drove out to Ma's house alone. At first I didn't go inside. I sat in the backyard, absorbing the stillness. Everyone was gone now but me. The only sound came from the bees buzzing among the scraggly roses. No kids did cannonballs into the pool, nobody played the guitar or poured another beer; Ma was not stalking through the garden, weeding between tomato plants. "No one lives there anymore," I said out loud. Caregiving had threatened to go on forever, but death was so shockingly final, so unbearably static.

For the next three weeks I felt like I was walking around in a world that belonged to everyone else but me; other people understood and functioned and communicated, while I dwelled in a separate compartment, underwater, isolated from other humans. Watching people fill their carts at the supermarket, I couldn't understand why they had to buy so many things. Things were just . . . things. Pat's silly humor, which had never failed to make me laugh, now made no sense to me at all, and I just stared at him, trying to muster even a feeble smile. I knew it must be funny; I just didn't speak funny anymore. No one else seemed to feel the immensity of Ma's absence, and that made me feel somehow invisible. At night, as I carried the recycling bins out to the curb, I looked up at the dark sky and found familiar patterns, the Big Dipper and Orion, and my heart stung with the familiarity of it. You could always find those stars, even on a brightly moonlit night, even midcity, but you couldn't find my mom.

Then death took over our backyard in Berkeley. Over the months I had spent focused entirely on Ma and the girls, something strange had happened in our back garden. It was a big yard, full of trees, and I was used to clearing away the yard waste every other month or so. Our two redwoods, deep green and peaceful and whispering tall in the sky, defined the yard, gave me a feeling of permanence and dignity. I respected the old trees and dutifully cleaned up after them.

But I was so focused on taking care of my family, I failed to notice that as the weeks and months rolled by, not only had far more of their needles begun to accumulate on the children's playthings, but also twigs and even small branches. Now, as I eyed the towering trees, I realized with growing dread that they had undergone a disturbing transformation. I saw beetle holes and fungus; they were leaking sap here and there, and their bark had turned from deep, moist, dark red-brown to the dry, grayish color of an old weathered barn. The ivy, which had previously left them alone, was creeping up into their lower branches, and now larger limbs in increasing numbers began to fall.

When my friend Doug, a botanist at Cal, offered to take a look at the trees, I felt hopeful; Doug might find a cure, Doug might save them. Instead, he told me the trees were very sick. They seemed to have botryospheria, a fungus that usually took hold in trees already stressed by another condition. "These trees are about as stressed as I've ever seen redwoods," he said. "See how they're making just hundreds of extra cones? That's a bad sign, too. It's like a last-ditch effort on the part of the tree to reproduce before it dies." He went on to explain that the original stressor may well have been sudden oak death, the mysterious blight also known as *Phytophthora ramorum* that was striking down hundreds of beautiful old trees in the region.

Since my walk with Ma in Briones Park six years earlier, I'd been aware of sudden oak death, and recently I'd seen it first hand, as certain trees gradually browned among the clumps of green. I feared that as these beautiful old trees fell in greater numbers, the familiar landscape of my childhood would be dramatically altered. Doug wasn't surprised that the infusion of vitamins and minerals our landlady had insisted on back in May had made no difference at all; there was still no known cure for *P. ramorum,* if it was the culprit. "The only way to really diagnose it would be to take the tree down," he said, "and even then sometimes the fungus doesn't leave a trace. But sudden oak or not, I'm pretty sure these two will die." It was difficult to predict how long it would take—three years, or even seven, he told me—but I was right to be concerned, because eventually the dying trees might fall.

I shook my head, staring down at the ground, wondering at the familiarity of that prognosis. He might as well have said we needed an autopsy. I surveyed the hundreds of little cones on the ground and thought of the way the brain in early Alzheimer's disease produces extra neurons, as if in anticipation of the damage to come.

Doug thought it was all about the trees. "I'm sorry, Syb. I know those were some handsome trees, but you should probably convince your landlord to have them taken out."

The day they took them, I was out all morning. When I returned home, as I rounded the corner I looked up as I always did to see those familiar shapes rising above our tile roof, but the scene was changed. The smaller tree was still intact, but the upper third of the larger one stood bare and blunt against the sky, stripped of its top foliage, the branches along its remaining trunk falling away as I watched. Men hung from ropes in the great tree's arms, and an enormous crane reached across our house to pull the branches away to the street. I stood in awe as the great crane slowly swung to street level and lowered two enormous branches to the ground. Next trip it brought a twenty-foot length of trunk. Slowly, the workers guided it down to lie in the long, dark truck bed.

Involuntarily, I suddenly envisioned Ma's body, her solid and familiar shape clad in her favorite red flannel shirt that she wore far too often, her motionless blue eyes still staring, still full of her Self as the mortuary men lifted her onto the gurney and zipped her into a black plastic body bag. I swallowed and pressed my lips together, looked up again at the man dangling at the top of our tree, and silently argued with myself:

They're killing it! one part of me insisted.

It's already dead, came the rational response.

They're taking it!

But we'll be safe now.

The problem was that we wouldn't be safe. We'd be safe from those two trees, but the whole wide world would remain. And Ma was dead, her body was dead, and they had taken it away.

The two people who managed to pull me back into the dynamic, breathing, laughing world every time I was with them were my girls. The primal imperative to be on call for them, to give them fresh, warm food and hugs, to give them my full attention and mama love, cut through the dull, rubbery curtain of grief that kept drawing itself around me.

Two-year-old Cleo still didn't get it about death. Looking at a recent picture of Ma, she said, "That's my grandmother who died."

"You called her Gram," I reminded her, "right, sweetie?"

"Yeah, and Gram is my grandmother. And she died."

"Yes, Gram died."

"Mama?"

"Yes?"

"Could Gram die back to her house now?"

At first Zoë was much less forthcoming, and I longed to read her heart, to see what Ma's death was doing to her sense of permanence and stability. I thought she understood that Gram was gone for good, but was she beginning to struggle with the further implication that any one of us could go at any time?

I was almost exactly Zoë's age the summer a car hit our cat Tabitha, the first time I really considered death. One morning there she was, dead in the street, with her eyes bulging out of her head. I didn't want to look at her eyes as my dad picked her up and carried her back to our yard. I felt guilty about that—as though I were betraying my cat friend by refusing to look at her.

For some reason my parents didn't bury Tabitha immediately. Daddy carried her down to the backyard and left her lying curled in the sun as she had often done, under the rosebushes beside the back fence. Seeing her there, I wanted to go say good-bye.

"Can I go see Tabitha?" I asked. Ma hesitated, but my dad said, "Let her go." Pragmatist. Atheist. Daddy believed this was a way to teach a child. It was, absolutely. The lesson would be deep and enduring.

"Is it okay if I touch her?"

"Yes," said Daddy, "but you should probably wash your hands afterward."

Ma made a face.

I walked down past the apricot and the orange trees, across clover buzzing with bees, to where Tabitha lay next to a garden shovel in the prickly grass. She'd been a full-grown cat but still young and playful. Her head was down and facing away from me, her body curled up as though for a nap.

Her sleek striped fur, warm in the morning sun, felt the same as it always had when I had pet her. If I didn't look at her face, I could imagine that she was still breathing. As I stroked her warm back and flank, my mind filled with the feeling that she was still alive, but when she didn't wake and rise to my touch, I knew that she was dead. I stilled, fascinated by this new experience, prickling with what felt like a secret, special understanding of life. I glanced up at the house, and then stroked Tabitha more firmly. Again, I felt the warmth of the sun, but this time also the uncanny heaviness in her neck and legs. I stopped, a little spooked. What if she woke up, dead? I had thoughts like that at age six.

Just as I had more recently with Ma, I stayed a long time with Tabitha, taking in her stillness. I cried a little, knowing I would miss her company, but more than her particular death, I absorbed the concept of death itself.

Now, when I thought of Tabitha's death, I imagined the millions of individual cells of her body blinking out like city lights when a transformer blows, *plink, plink, plink.* As a child I hadn't known what a cell was, that we are all made out of millions of microscopic building blocks, each one with a set of instructions for the manufacture of goods and a set of tasks to carry out in life. I hadn't known that blood cells move through vessels, carrying oxygen to, and carbon dioxide away from, every cell, or that stationary cells in the kidney helped to collect and deposit waste. I hadn't known that nerve cells communicate messages all over the body and brain by conducting electricity down their long, skinny arms, or that when the heart stops beating, its oxygen-deprived cells send out a message that signals a massive cellular suicide. All I had known was that Tabitha for sure was not coming back. All that was left of her was that body, which was about to be buried and decompose—that and our memory of her.

I got a hint as to Zoë's state of mind the week before Halloween. Her school librarian, Ms. Charleston, who always worked around seasonal themes, had set out the October selection, and Zoë came home with a book by Gina Freschet called *Beto and the Bone Dance,* about Mexico's Day of the Dead. "I got this for you, Mama," she announced, "you know—because of Gram." We snuggled on the couch and read about a magical night in Mexico, and a boy who wasn't sure what to bring to his grandma's grave for the celebration of El Dia de los Muertos, until he realized the best treat for her would be a picture of him.

The next day I drove through the rainy streets of Oakland to the crematorium and collected Ma's ashes: a full tin box, heavy inside a green faux-velvet bag. We set the box on a small wooden table in the living room, and around it we built an altar, with photos of Ma at different ages, and of Ma's favorite people: Daddy at the dining room table, holding his guitar and mug of ale; Eddie, smiling crookedly at the camera through thick horn-rimmed glasses; Grandma and Gramps on the front porch of their little family home in Orange County; Alice and me as kids, and then again as adults with our husbands and the four grandchildren; and Ma's best friends, Vicky, Daniel, and Eleanor. Behind the pictures, I stacked books by some of her favorite authors—Louise Erdrich, Barbara Kingsolver, Michael Dorris—and a bumper sticker from her beloved lefty radio station, KPFA. I filled a ceramic demitasse with strong black coffee, which I set next to a dish of the Hershey's Halloween candies she always bought and a little sugared loaf of Day of the Dead bread in the shape of a turtle.

All this we surrounded by several candles, a handful of smooth ocean rocks, and five baby sand dollars from Stinson Beach. We lined the table's edges with the bright orange heads of fresh miniature marigolds, and then Zoë helped me make a thick golden path of marigold petals running from the front door to the altar table, so Ma's spirit could find her way in. On November 2 we ate a special Cheese Board pizza and salad in remembrance of Ma, with Bread of the Dead and Halloween candy for dessert; that night at bedtime, instead of reading from the girls' books, we each told a story about Ma. For weeks after that, every night I lit candles for Ma before dinner, and the kids blew them out before bed.

By the time Alice and her family arrived for the holidays, the altar had taken up semipermanent residency in our living room. The marigolds and food were gone, and I had moved the photos to another shelf, but the girls continued to collect treasures and present them to Ma: pebbles, shells, flowers, and twigs surrounded her tin of ashes.

Ma's memorial was sweet and informal, just like her. We gathered at my house, ate, drank, and traded stories. Alice and her family were there; Daniel and Laurel came, and a few more of my friends. Ma's best friend from work, Vicky, whom I'd thought of as my keynote speaker, got lost on the thirty-mile drive from Pittsburg. She called late that afternoon. "I hate to say it," she said, "but I think I might have a touch of Alzheimer's myself. This has been happening more lately—I'm so sorry. I loved Ruthie so much, Sybil."

We each wrote Ma a note. I instructed the small group to write anything—good wishes, farewells, forgiveness, or a memory. I watched Zoë carefully pen her note with a red felt-tip marker, in blocky kindergarten letters. Alice wiped away tears, taking her time, and deposited three different messages into the pot. Daniel's note took no time at all. We took them all outside and set fire to them in one of Ma's old iron skillets, sending our thoughts out into the universe at large, the way I imagined Ma had gone when she had been cremated. Fog wafted up the hill from the bay, and it felt good to be outside.

CHAPTER 21

Writing Home

One morning that February as the kids ate their oatmeal and I pared kiwis for their lunches, Pat called from a BART train on his way to an off-site meeting. Unsure what station he needed, he asked me to look up an address in San Francisco, but after I gave it to him, his train entered the transbay tunnel and we got cut off. A few minutes later, the phone rang again.

"Hello?" I said impatiently, expecting Pat again.

But it was Ma.

"Sybil?" she asked nervously. She had a worried, searching way on the phone, a sort of pleading, half-panicked tone. I heard the strain of it. She hated those first moments after she dialed, listening to the ring before I picked up. Something about reaching out into the dark emptiness of the electrical connection was very stressful for her.

I hurried to put her at ease. "Sure, it's me," I soothed. "Are you okay?"

And then in rushed the confusion, the knowledge, the dissonance: she's dead.

"Sybil, it's Norah."

"Norah?" The noise in my head was so loud, I almost couldn't hear the words. Not Pat, not Ma; my friend Norah, a local poet. I hadn't heard from her in weeks.

"Yes." She laughed nervously. "I know this is unexpected. I hope it's not too early to call."

I somehow managed a short conversation with Norah, who asked twice whether I was feeling okay before I excused myself. I couldn't explain it to her. It was almost nine. I had to get the kids into the car, or Zoë would be late for school. Cleo had her sandals on the wrong feet, and Zoë wanted more milk. After I hung up, I held the phone to my heart. Ma had been there, on the line, her voice in my ear as I stood in my kitchen. My eyes felt hot, and my breath was gone. I knelt on the kitchen floor, took off my glasses, and covered my face, breathing in the scent of my lotion as the tears came. It wasn't just me thinking someone sounded like her. It was Ma. I had spoken to her.

Zoë was standing in front of me.

"What happened, Mama?"

"Oh my God," my small voice said. I was so overwhelmed, so lost in the moment, that it didn't occur to me to filter or soften my experience for her.

"What, Mama?"

"That was Gram . . . I mean, I heard Gram's voice."

Zoë tilted her head, waiting to understand.

"I miss her, Zoë. I miss Gram. When I heard that voice, for a second I thought . . ."

She stepped in close and carefully hugged me. Zoë has an instinct for the important moments in life, a way of taking me seriously when it matters. For a moment she held her soft cheek still against mine, comforting me the same way I sometimes did her.

"I miss her, too, sometimes, Mama."

And as I relaxed into Zoë's embrace, I realized that what I was feeling was relief. It had just felt so good to hear Ma's voice. She had called me. She was there.

That morning seemed to initiate a several-months-long process of emotional extremes. The moderate me seemed to have been excised, leaving me open, raw, and sensitive. At certain moments I turned away from other people, and a shadow fell across me. Heavy and slow, I would fill with the dull thud of futility: *So what, so what, we're all just going to die.* Then suddenly, as I piled the girls into the car in their pajamas for evening story time at the library, or walked with them under our umbrellas in the rain, everything would change; each moment would strike me as crisp and poignant, infused with meaning and direction.

One day Cleo and I encountered three large mutts outside our neighborhood café. As I stood sipping my latte and chatting with their person, Cleo respectfully approached one dog and asked him if he would like to be petted. She held out her hand for the sniffing. This formality dispensed with, she tentatively stroked the fur on his back, and then stood coyly smiling, holding her cheek invitingly in position for the dog to kiss. When he obliged, she let off a peal of such deeply gratifying giggles, everyone on the surrounding sidewalk turned and laughed with her.

As we walked on down the avenue, she said, "Mommy? I think I'd rather be a dog."

"Oh yeah?" I asked. "What do you think you'd like about that?"

"I'd like to know how it feels to chew up a bone."

"Oh yeah," I said, nodding. "Anything else?"

"Yeah: then I could have my own water bowl."

What a thought: *I think I'd rather be a dog. I would like to learn lithography,* I thought; *I'd like to see Alaska.* But *be a dog.* Cleo still saw birth into another species as a viable path in life.

Her openness to every fold of existence, the way she gave each possibility equal weight, made me giddy.

Pediatric psychologists have documented the simple yet supple way the brains of Cleo's young cohort work; they really do process experience differently from us. One difference is that preschoolers are less susceptible than older children and adults to the effects of context. For example, consider two identical circles, one of which is surrounded by a ring of smaller circles. When asked to assess the relative sizes of the naked circle and the surrounded circle, grown-ups tend to name the circle with an entourage as the larger of the two, whereas toddlers are far less susceptible to this illusion. This apparent disregard for visual context is paralleled in young children by the unencumbered way they interpret the present moment: toddlers tend not to make use of past mental images or knowledge of the world to the extent that older children and adults do. So at eight, Zoë, upon seeing a bucket of water on the back porch, would retrieve and integrate the information that it had rained for several days before and say, "Mom, look how much it rained!" But four-year-old Cleo, who had also experienced the rain, was much more likely to say, "Look, there's a bucket of water on the porch!" This tendency of the younger set not to view the world through the lens of past experience might help explain the creative ease with which Cleo considered life as a dog; she put much less emphasis on what she knew (she could be a dog only if Mommy and Daddy were dogs) than what she was experiencing presently (dogs were soft, fun, happy beings).

These differences in the way toddlers think, which drop off dramatically during the elementary school years, have been ascribed to a lack of integration in their young brains. Soon, as glial cells myelinized the axons connecting Cleo's various brain regions and these regions began to synchronize their neuronal activity, context and history would become more prominent in her thinking. One might view her present state as a limitation—Cleo wasn't taking all the facts into account—but it also afforded her the freedom to think outside of the contextual box. In a few years she would almost certainly no longer consider dogdom an option, but for now she was not going to rule it out.

I felt so static by comparison, so far from that time of possibility. The path kept narrowing. I could just hear the doors quietly closing as I began down the gentle slope of my forties. I would never be a surgeon or a track star; I probably wouldn't ever bobsled or blow glass. My habits—shlumpy clothes, too much coffee—were deeply entrenched. And worst of all, my once so stimulating relationship with my husband, tattered and aching after years of stress and half-acknowledged resentment, was now sliding into a familiar repeat loop of nitpicks and bad habits, threatening to devolve into mediocrity or worse. Some of the pressure was off now that Ma was gone, and we fought less frequently, yet we could go for days lying in bed next to each other, plugged in to our laptops, unruffled but also unengaged. A hand might stray over to say hello. We might giggle and relate an anecdote. But for the most part we remained in our separate worlds, me writing, him surfing the Internet.

And yet there was hope for us. It turns out that grown-up brains can evolve, too. There are certainly set neurobiological windows of opportunity—most famously, the language stage Cleo was in the midst of. Some of these windows all but close later in life; no matter how hard I try, I will never perfect the French accent I started working on in college. But as I had experienced in early motherhood, not only do synapses form and continue to be refined in response to experience throughout adulthood, but brand-new nerve cells are born into the brains of middle-aged grown-ups. And it turns out that these cells don't just play a role in the olfactory bulbs of new mothers. New neurons are also continually born in the dentate gyrus. This is where Pat and I might find hope: the dentate gyrus is a part of the brain that helps us to form new memories and may even regulate happiness. The adage that you can't teach an old dog new tricks is cute, but it may not be true.

Enter: the camera and the pen. That year Pat's big purchase had been a fancy digital camera with single-lens reflex. Since he'd bought it, he had been plaguing us all, like some kind of in-house paparazzo, popping up in the middle of every activity to snap shot after shot after shot. He'd even come down and recorded my two a.m. escapade into the garden to pick slugs off of our straggly green bean seedlings.

Meanwhile, I would slip off to a café for some alone time and find myself writing the story of how we met twelve years earlier, on opposing intramural volleyball teams composed of geeky graduate students. I smoothed out such memories, like letters that had been riding around in my pocket all these years, and read them back to myself with pleasure.

What were Pat and I doing with these preoccupations of ours—his visual, and mine verbal? Late one night I was writing a column for the online literary magazine my writer's group had started and he was fiddling with his new camera, when we both paused in the kitchen for a cup of tea. He grabbed his camera and started snapping away at me. I said, "Stop it, you're driving me nuts!" He replied simply, "But I just want to take pictures of you." There was an urgency and earnestness in that plea that stopped me. His voice was full of love—alive love, the fresh kind.

That was when I realized that he was finding a way to look at me, just me, through a frame that eliminated all the baggage and complication; me, the person he loved. And I had been writing about him in that same way; I sought to see him, the tip of my nib scratching black shapes onto the page. He purred, "Work it, baby!" as his shutter opened and shut with a satisfying Hollywood *shllk!* And I scolded him, but I smiled, because he annoyed me, but I loved him. I wanted him to see me. We managed to really look at each other, for an instant, to recapture each other's image.

Subtly, those few moments and other moments like them would initiate a change. There would be the slightest physiological shift. At moments like that, my thoughts became cool and soothing, and my pulse slowed. I imagined a little bump on a dendrite, a tiny bit more serotonin or dopamine crossing a synapse; a change of mind. I could see the neural pathways that had formed in our first days of flirting on the volleyball court, reactivated now, as we stood in our kitchen laughing; the tips of our neurons reshaping themselves ever so slightly, bringing delicate changes in our perceptions of each other. Adjustments. New tricks.

As we get older, our mental processes, like our physical ones, slow down. But it turns out that when it comes to memory tests, healthy seniors can perform just as well as young adults, provided they are simply given more time to respond. It seems older people need a slower world, not necessarily a simpler one. I liked this idea of clear-minded deceleration. I liked the thought that Pat and I might manage to slow down together. *So let Cleo and Zoë ramp up into the frenzy of their young adult life,* I thought. *Let them experience the massive formation, pruning, and reshuffling of their billions of synapses.* With Pat and me, it was bound to be a more subtle progression from here on out, a refinement, a smoothing and tending affair. Things had not been perfect for us; I knew we had a tremendous amount of work ahead. It was a good thing we were both stubborn as hell.

•   •   •

When Ma died, she left her house to Alice and me, but as executor of the will, I wasn't sure just what to do with it. People advised me to keep it; this was the San Francisco Bay Area, where real estate prices never seemed to stop climbing. They suggested that Pat and I occupy it, to save money, and just watch our nest egg grow. But I couldn't imagine returning to my little hometown, surrounded by oil refineries and a forty-minute drive from the Berkeley community I loved; I couldn't imagine living in that funny little place, nine hundred square feet divided into two separate floors with no inner stairs. I thought about buying my sister out and fixing it up—or rather, going into debt paying somebody else to fix it up—and becoming a property manager. Then I remembered insurance and taxes and liabilities and bad renters. Suppose I had to evict somebody. I'm not the evicting type. It made much more sense to sell it. My sister agreed, and so did Pat. We even found a buyer, a friend of Daniel's in town, who was recently divorced and looking for a pet project. This guy actually liked our strange little fixer of a house. It made sense to sell to him.

The problem, which nobody but I could see, was that I had to weigh that reasoning against the cool morning air, a soft breeze off the Carquinez Straits, Ma's roses along the fence. How could a home price index capture the worth of red-wing blackbirds in the fruit trees, a ripple in the pool, or the thump of a green apple plopping into the dry grass? I had to weigh logic against the familiar crack in the concrete side steps, where I'd learned to sweep and my mother had praised my hard work when I was ten; the fern I had rooted in a glass of water and planted next to those steps when I was twelve, that had now grown almost into a hedge; the big white peace sign Ma had painted in the center pane of the front windows decades ago in acrylic paint, still there providing a testament to her loving spirit. Sitting with all this just felt more important to me than obtaining appraisals and comps, than calling my real estate friend for help with a sales contract.

So I sat on it. I brought Cleo and her dolls, my Peet's coffee, the probate forms, and my Nolo Press book of legal advice, and I sat *in* it. I watered the yard and checked on the pool and, wandering from room to dirty room, peeling up a corner of carpet here, a scrap of wallpaper there, I remembered. Again and again I returned to a particular patch of old wallpaper high in the small room that had once been the kids' room, though it had doubled as Ma's library and bedroom for several decades since; the paint had peeled away to reveal the pastel blue and soft pink stripes of our nursery. Seeing those colors in that particular pattern pulled at my heart, and I sought out that feeling again and again.

That wallpaper had been covered up by paint early in my childhood, yet it had managed to thoroughly permeate my consciousness; seeing it took me places there was no other way to get to. The strength of such a memory depends on several factors. Three of the most important are repetition, emotional content, and meaning; if an event is highly meaningful, emotionally charged, or repeated over time, it is more likely to be retained in long-term storage in the brain. It may not be perfectly accurate, but it is likely to be strong. So no wonder I strongly associated that wallpaper with feelings of safety, of protection, of Mommy and Daddy being nearby. No wonder it seemed so very familiar; highly meaningful, emotionally charged, and often seen, it had been multiply reinforced—and here I was making those connections even stronger for myself by revisiting it now; I imagined myself toiling away to lay down the dendritic asphalt—I had to reinforce that cellular highway back to my childhood before I left the premises.

From time to time Alice called from Seattle. "How's it going with the house?" she'd ask. I'd say, "Oh, slowly. It's hard to get a lot done, what with the kids and all, but I filled out the waiver of account and notice of hearing. You'll get some papers to sign next week."

It would have been harder to tell her, "I cried all morning, sitting beside that Meyer lemon tree." I wouldn't get to bring the kids to pick fruit at Gram's house anymore, once I sold it. With no more visits there, they would forget her, forget why we went to that place, forget what used to be inside. I had photographed all the titles of my parents' hundreds of books before I sold them, and played a slide show of the jpegs on my computer at home as a screen saver. (Somehow a couple of shots of Cleo in the buff got in there. Interspersed with volumes of T. S. Eliot, *Pogo,* Mary Stewart, James Thurber, e. e. cummings, John Steinbeck, Anaïs Nin, and Charles Dickens, every sixth picture or so a naked, mischievous toddler flashed by.) *Maybe,* I thought, *I should also photograph the wallpaper, the cracks in the side step, and the place in the back of Ma's closet where one of us kids had clumsily scrawled the F-word about a million years ago.*

But what would be the use of that? My kids would inherit boxes full of photos—along with the boxes of unlabeled photos I dragged home from Ma's house, some of which she shipped up from her own mother's house in the mid-1980s, during that probate case. Zoë and Cleo would probably feel the same helplessness and despair I now do—it's alarming, really, when you connect your own blood to the mysteriously familiar faces in those old photos. No, not even that old—but already forgotten, already unknown. Piles of bones upon piles of bones, and the only meaningful record of them now is not in the nameless photographs but in the nuclei of my daughters' cells, the genetic pattern of our ancestors carried within them, unconsciously affecting them, but never explicitly. I wanted to hold on to all these details because letting go meant that I, too, would be forgotten.

Pat and I occasionally entertained the fantasy of ownership, of giving our children a sense of permanence: *Here is our spot. You may always return here.* A house gives the illusion of stability, this feeling of home. But in truth you may not return. You aren't the same *you* any longer, and even if you were, the place you return to always changes. The people who occupy it have aged. The lovely times or sad times you spent there are past. The new next-door neighbors have cut down their ash tree and planted oleander in its place. We regard Place as a door to the attic where our memories are stored; open the door, retrieve the memories. But even that is a partial fallacy, because the memories themselves change. Every time we root around in the attic, we stir things up, removing some things and adding others; we revise our memories, adding to them the context of our present lives. I can clearly recall the wallpaper of my childhood nursery, but now it is inextricably tied to my mother's death, and to the empty, quiet house I had to sell, in addition to that warm and protected room of my childhood. Subtle but tangible changes in the fine detail of our synaptic architecture are made in response to the pattern of electrical activation in our brains, making it physically impossible to retrieve the exact original memory.

Regardless of our plans for the house, I had to clear out Ma's belongings. I sorted through box after box of mildewed clothes and broken dishes; I threw out decrepit cassette tapes of radio KPFA's grassroots social activism program *Democracy Now* and filled an entire shoe box with toenail clippers, files, tiny scissors, and queerly bent metal picks, all of which had hung on a magnetized bar in the bathroom; I sold a truckload of books to Black Oak Books, our favorite local independent bookstore.

When everything truly valuable or truly worthless was gone, I held an estate sale. I stood in the front yard of the house I had grown up in and watched as strangers carried the Lockharts away, piece by piece. It was a cool day, but the doors at Lafayette Street were flung wide; the yard bustled with eager bargain-hunters, and I stood in a daze. There went our music, our books, our tools, and our furniture, walking off down the street. As all the little pieces of our life scattered, I felt my center slipping away; the switch had been thrown, and I was back in the darkness.

Over coffee I told my friend Polly how futile life felt. "People just die," I said, "and we sell their stuff. Why do we worry so much about everything—education, exercise, morals? Why bother?"

Polly had called me often in those first weeks. "Just checking in," her voice had said to my answering machine, day after day. Some days I smiled and pressed three to delete, and other days I stood there, tears streaming down my face, struck by the depth of her simple, persistent offer. "I love you," she said. "I'm here." Now she looked at me carefully and said, "Maybe the best reason to keep going is just to keep each other company while we're here." I thought about that, and I decided that she was right. I wrote it in my journal that night, and as I did, I realized that this is also the reason why people write: to keep each other company.

The next day one of my mom's neighbors called me. She'd bought a desk at the estate sale, and in the bottom drawer she'd found a folder she thought I might want. She told me it seemed to contain a collection of Ma's notes and letters from the fall of 1992. That was four years before we had begun to notice the change in her.

I sat in our kitchen, which was bright with unfiltered sunlight now that the redwoods were gone, and read through thirty-seven pages, handwritten on thin-ruled notebook paper, in Ma's neat schoolteacherly hand. Some were letters to authors:

Dear Barbara Kingsolver, one letter began, *I feel like starting this letter, "Dear Alice." That's because after finishing* Pigs in Heaven, *Alice is real to me, like a friend.*

There were letters to President-elect Bill Clinton, welcoming him to the presidency, to L.L. Bean, thanking them for their excellent rain gear, to Amnesty International regarding her yearly donation, and to air board member supervisor Tom Powers, demanding that he better inform the community of refinery and chemical plant safety and air emissions. There was a handwritten note from author Annie Dillard, thanking Ma for her kind letter regarding her book *The Living*. And then there were what seemed to be journal entries, like this one:

Sept. 21

I need to write even though I don't know what to write about.

I need to keep lists, learn how to remember people's names better, and have better ways of keeping things neat. This place is a mess.

I need to think of good things and live people.

I need to not voice criticisms, to be quiet or say positive things.

I need to do more exercise, or at least more walking.

I forget too many things. I forgot to ask for another prescription for the lotion I need to put on my arms to prevent tingling at night. I forgot to take the film to Valco to be developed, even though I put it on a list. I forgot the list. So maybe that sounds funny, but I worry about forgetting things.

I just forgot the word "cortisone," but I'd written it down and found it quickly.

I can remember Grant's wife's name Trini.

I don't think I want winter to come.

Ma had written that when she was sixty-two years old, five years before my return to California and almost exactly ten years before she died. As I read, it occurred to me that this writing from the time of her very earliest symptoms was a sort of trail marker for me, allowing me to reconnect in reverse order to that mother I loved. Just as she had begun to lose her memory, she had laid down this record, and now I used it to trace my way back to the whole Ma, Ma who read ardently and passed her books to me (she gave me *Pigs in Heaven* that year; I remembered reading it at Stinson Beach) and rallied for causes, Ma who always turned her warmest, kindest side to the world, keeping her darkness to herself. I wondered whether letting that darkness out more would have been better for her health, but it was pointless to dwell upon that now.

In Alzheimer's disease, myelin is lost in the reverse order of its appearance during development. This means that the last abilities babies acquire are the first to be lost when an adult has Alzheimer's. The consolidation of memory from short- to long-term, for example, is slow to develop in babies, because the hippocampus is one of the last structures to become myelinized; this same ability begins to wane early on in Alzheimer's disease, since it is one of the first to lose its myelin. This reverse-developmental progression is one reason why Alzheimer's patients seem more childlike as the disease advances. I thought about this when I realized what I was reading. For the past few years I had been submerged in the daily degradation of Ma's personality, but since her death I had begun to search for the memory of her wholeness. Now, with these writings, I was regaining that wholeness in the reverse order of its disappearance. This last bit of evidence of her active mind dated from the time when she had just begun to doubt her memory, and I had gained the desire to write just after she had lost it.

It scared me a little, having watched Ma disconnect from her world synapse by synapse, to rediscover her here, reaching out from the past with these words, trying to remain connected in just the way I had begun to do these years later. Was I subconsciously doing the same thing, collecting my memories up into columns and stories because I intuitively knew they would be lost when I, too, began to experience the symptoms of Alzheimer's? This was a fear that would not leave me anytime soon. My main consolation was that writing was thinking, and thinking was good for my brain.

More than anything, Ma's journal reminded me of the connecting power of this most solitary and yet sociable pastime. The writer, whether a famous author or an ailing grandma, cut a path; the reader walked down it, and before the writer had even finished the scene, the reader would find a solution, have an emotion, resolve a dim memory, or simply discover something within herself—a strength or creative force that rushed out to meet the writer, extending the path in collaboration. Words connected me to my children as they grew, and to my mother as she faded away. They connected me to myself. And they reattached me to the world I lived in. I realized that part of why I wrote was to let Ma's story out into the world, where it could be held and nurtured by other people's understanding; I wanted to write my mother home.

A week after the yard sale, I hired a local guy to do dump runs, which he did until the house was bare. Six months after Ma's death, when Daniel's friend Mark called to inquire about the property, I worried whether I was being a responsible custodian of her affairs. Wasn't I losing money by keeping that house empty so long? Could I get more for the estate if I went through the sleazy real estate agent down the street who had approached me at the yard sale and dogged me ever since? But still I made vague excuses to Mark, and I sat with it a little bit longer; I watered the garden every week and wrote, and cried. I cried as I drove those familiar roads, bronzed hills dotted with black-and-white dairy cows. I cried to and fro, slowly cleaned the place up, and did a bit of paperwork here and there on Ma's probate case. Finally, after one full year, I bought yet another Nolo Press book—this time *For Sale by Owner in California*. I gathered the comps, got the appraisal, asked Mark for his offer, and completed the relatively uncomplicated forms required to sell him the house. Acceptance is the flip side of fear. I had to sit there and love my family, love my past, absorb the feeling of that home. It had taken a year of weekly visits to come to peace.

One sunny Monday morning in August 2005, I went to the old county courthouse in downtown Martinez for the last time. Glancing at the low office windows along the side street, I remembered how my friend Heather and I used to hide behind the bushes and "spy" on the county workers in there at their desks when we were ten. Feeling like a child in disguise, but looking for all the world like a normal middle-aged grown-up, I walked straight up the brick steps and through the glass front doors, surrendering my bag to a police officer for inspection before I proceeded to the courtroom. Inside, from a seat set above me and the other citizens he served, an honorable commissioner gave me permission to officially close Ma's probate case.

After filing the form with his signature on it, I drove past the house one last time. I was intending to merely peer in like a stalker from the safety of my car, but Mark, the new owner, was there laying bricks in the front yard and invited me in for a tour. Inside, for a moment I envisioned Alice and me, running through the rooms, and superimposed upon us, Zoë and Cleo, all of us bearing such a strong likeness to my mother, and hers. Then I took in the changes: Mark had scraped off all the old wallpaper and repainted; he had replaced the front roses and scraggly weeds with a lush green lawn; and the oak tree I had received as a high school graduation gift, which had grown to dominate the front yard, was gone. Light streamed into the once darkened and impossibly grimy house; the floors gleamed, and the fresh morning air flowed easily through the rooms. The house no longer contained us, but it had come alive. Mark told me, "I'm replacing the front windows, but I'm saving that peace sign. And you can come pick the Meyer lemons anytime; I'll give you lemon rights for life." That was when I knew I had chosen the right buyer.

In a sense, as I had walked out of the courthouse that afternoon, I had almost finished putting Ma to rest—at least Paper Ma, the entity she had become: a court case number I'd typed into so many downloaded forms, it had become as familiar as a surname to me. And I was still afraid, even though the house was in good hands, that Ma and Daddy and all our life there would lose its shape irretrievably now that I wouldn't have glimpses of wallpaper and the smell of the garden to coax the memories out of me. I feared Ma would become two-dimensional, like the girl in *The Tall Book of Make-Believe* whose brother pulled her under a door, leaving her as thin as a page.

I needn't have worried. Ma is not a document in my accordion case file. Every month or so she emerges, the way she did that morning in my kitchen when she "called" me. Sometimes she comes bumbling and confused, as she was in her last days; others she appears as her earlier self, confident, generous, and comforting. She arrives without any particular agenda, just making her presence known.

Friends have told me that they came to believe in ghosts or spirits only after the loss of a loved one, when they began receiving such visits. I think people underestimate the power of memory, which to me is a personal creation, just the way writing is, except that what we create is a living, physiological embodiment of the person who has passed, encoded in elaborate mazes of synapses that have formed and re-formed over a lifetime shared with another person. Our loved ones literally never leave us. And we incorporate parts of ourselves within our memories just as we do in composing a poem; we merge with other people in this way, and out of the merging bloom memories so fresh we can almost touch them.

One night as I sat writing at my desk, about a month after I'd sold her house, Ma appeared, utterly vibrant, on the street below my study, arms folded, looking up at me full of fierce and protective mother-love. This was the long-ago young and capable Ma, the one who'd been in charge of Alice and me when we were little. As I gazed down at her, I heard Zoë stir and mumble in her sleep, and I had the urge to run and wake her and Cleo. *Look!* I would say. *Gram has come to visit!* I wanted them to know this version of my ma. But I didn't turn away. I noticed the slight flex in Ma's calf muscle as she stood rooted to that spot below me. I saw her eyes reflecting the light from my window, and she pushed back a strand of hair, such a familiar gesture. She stirred, in the unique way of my ma, and she disappeared again.